T0255951

SpringerBriefs in Applied Sciences and Technology

Forensic and Medical Bioinformatics

Series editors

Amit Kumar, Hyderabad, Telangana, India
Allam Appa Rao, AIMSCS, Hyderabad, India

More information about this series at http://www.springer.com/series/11910

Ch. Satyanarayana · Kunjam Nageswara Rao
Richard G. Bush

Computational Intelligence and Big Data Analytics

Applications in Bioinformatics

 Springer

Ch. Satyanarayana
Department of Computer Science
and Engineering
Jawaharlal Nehru Technological
University
Kakinada, Andhra Pradesh, India

Kunjam Nageswara Rao
Department of Computer Science
and Systems Engineering
Andhra University
Visakhapatnam, Andhra Pradesh, India

Richard G. Bush
College of Information
Technology
Baker College
Flint, MI, USA

ISSN 2191-530X   ISSN 2191-5318 (electronic)
SpringerBriefs in Applied Sciences and Technology
ISSN 2196-8845   ISSN 2196-8853 (electronic)
SpringerBriefs in Forensic and Medical Bioinformatics
ISBN 978-981-13-0543-6   ISBN 978-981-13-0544-3 (eBook)
https://doi.org/10.1007/978-981-13-0544-3

Library of Congress Control Number: 2018949342

This Springer imprint is published by the registered company Springer Nature Singapore Pte Ltd.
The registered company address is: 152 Beach Road, #21-01/04 Gateway East, Singapore 189721,
Singapore

Contents

Chapter 1
A Novel Level-Based DNA Security Algorithm Using DNA Codons

Bharathi Devi Patnala and R. Kiran Kumar

Abstract Providing security to the information has become more prominent due to the extensive usage of the Internet. The risk of storing the data has become a serious problem as the numbers of threats have increased with the growth of the emerging technologies. To overcome this problem, it is essential to encrypt the information before sending it to the communication channels to display it as a code. The silicon computers may be replaced by DNA computers in the near future as it is believed that DNA computers can store the entire information of the world in few grams of DNA. Hence, researchers attributed much of their work in DNA computing. One of the new and emerging fields of DNA computing is DNA cryptography which plays a vital role. In this paper, we proposed a DNA-based security algorithm using DNA Codons. This algorithm uses substitution method in which the substitution is done based on the Lookup table which contains the DNA Codons and their corresponding equivalent alphabet values. This table is randomly arranged, and it can be transmitted to the receiver through the secure media. The central idea of DNA molecules is to store information for long term. The test results proved that it is more powerful and reliable than the existing algorithms.

Keywords Encryption · Decryption · Cryptography · DNA Codons · DNA cryptography · DNA strand

1.1 Introduction

DNA computing is introduced by Leonard Adleman, University of Southern California, in the year 1994. He explained how to solve the mathematical complex problem Hamiltonian path using DNA computing in lesser time [1]. He envisioned the use of DNA computing for any type of computational problems that require a massive amount of parallel computing. Later, Gehani et al. introduced a concept of DNA-based cryptography which will be used in the coming era [2]. DNA cryptography is one of the rapidly emerging technologies that works on concepts of DNA computing. DNA is used to store and transmit the data. DNA computing in the fields of

© The Author(s) 2019
Ch. Satyanarayana et al., *Computational Intelligence and Big Data Analytics*,
SpringerBriefs in Forensic and Medical Bioinformatics,
https://doi.org/10.1007/978-981-13-0544-3_1

Table 1.1 DNA table

Bases	Gray coding
A	00
G	01
C	10
T	11

cryptography and steganography has been identified as a latest technology that may create a new hope for unbreakable algorithms [3].

The study of DNA cryptography is based on DNA and one-time pads, a type of encryption that, if used correctly, is virtually impossible to crack [4]. Many traditional algorithms like DES, IDEA, and AES are used for data encryption and decryption to achieve a very high level of security. However, a high quantum of investigation is deployed to find the key values that are required by buoyant factorization of large prime numbers and the elliptic cryptography curve problem [5]. Deoxyribonucleic acid (DNA) contains all genetic instructions used for development and functioning of each living organisms and few viruses. DNA strand is a long polymer of millions of linked nucleotides. It contains four nucleotide bases named as Adanine (A), Cytosine (C), Glynase (G), and Thymine (T). To store this information, two bits are enough for each nucleotide. The entire information will be stored in the form of nucleotides. These nucleotides are paired with each other in double DNA strand. The Adanine is paired with Thymine, i.e., A with T, and the Cytosine is paired with Glynase, i.e., C with G.

1.2   Related Work

There are a number of existing algorithms in which traditional cryptography techniques are used to convert the plaintext message into a DNA strand. The idea of DNA which is a type of encryption, if imposed exactly, is virtually uncrackable if applied in the molecular cryptography systems based on DNA and one-time pads. There are various procedures for DNA one-time pad encryption schemes [1]. Popovici [4] proposed a cryptography method using RSA algorithm. He simply converted the plaintext into binary data and converted the binary data into its equivalent DNA strand. He used RSA algorithm for key generation. Yamuna et al. [7] proposed a DNA steganography method based on four levels of security using a binary conversion table. Nagaraju et al. [8] proposed another method for level-based security which provides higher security rather than the method proposed by Yamuna et al. In the DNA strand, we use only four letters, so there is a possibility of hacking the information. To avoid this, the following algorithm is proposed which uses DNA Codons. Hence, by choosing any three letters of DNA strand, we can form 64 combinations of Codons represented in Table 1.2 [9]. Out of these 64 Codons, 61 Codons form 20 amino acids and 3 are called as stop Codons which are used in protein formation

Table 1.2 Structured DNA Codons [9]

second base in codon

first base in codon		T	C	A	G	third base in codon
T		TTT Phe	TCT Ser	TAT Tyr	TGT Cys	T
		TTC Phe	TCC Ser	TAC Tyr	TGC Cys	C
		TTA Leu	TCA Ser	TAA stop	TGA stop	A
		TTG Leu	TCG Ser	TAG stop	TGG Trp	G
C		CTT Leu	CCT Pro	CAT His	CGT Arg	T
		CTC Leu	CCC Pro	CAC His	CGC Arg	C
		CTA Leu	CCA Pro	CAA Gln	CGA Arg	A
		CTG Leu	CCG Pro	CAG Gln	CGG Arg	G
A		ATT Ile	ACT Thr	AAT Asn	AGT Ser	T
		ATC Ile	ACC Thr	AAC Asn	AGC Ser	C
		ATA Ile	ACA Thr	AAA Lys	AGA Arg	A
		ATG Met	ACG Thr	AAG Lys	AGG Arg	G
G		GTT Val	GCT Ala	GAT Asp	GGT Gly	T
		GTC Val	GCC Ala	GAC Asp	GGC Gly	C
		GTA Val	GCA Ala	GAA Glu	GGA Gly	A
		GTG Val	GCG Ala	GAG Glu	GGG Gly	G

[10]. This gives rise to ambiguity like Phenylalanine amino acid mapped on TTT and TTC. To overcome this, we prepared a Lookup table (Table 1.3) for each Codon. A Codon is a sequence of three adjacent nucleotides constituting the genetic code that specifies the insertion of an amino acid in a specific structural position in a polypeptide chain during the synthesis of proteins.

1.3   Proposed Algorithm

The above 64 Codons (Table 1.2) can be used to encrypt either text or image. In the present case, we propose an algorithm to encrypt text only. We want to encrypt the text that contains English uppercase or lowercase characters with 0–9 numbers including space and full stop that count 64 in total. The following Lookup table (Table 1.3) shows the Codon and its equivalent character or number that is going to be encrypted. In our algorithm, we implemented the encryption process in three levels only. As the number of levels increases, the security also increases. The main advantage of this algorithm is that the Lookup table gets arranged randomly each time the sender and the receiver communicates. As a result, the assignment of the character also changes every time which is a challenge to the eavesdropper to crack the ciphertext. This Lookup table is sent through a secure medium.

Table 1.3 Lookup table

S. No.	DNA Codon	Replaceable character	S. No.	DNA Codon	Replaceable character	S. No.	DNA Codon	Replaceable character
1	TTT	A	22	CCC	V	43	AAA	Q
2	TTC	B	23	CCA	W	44	AAG	R
3	TTA	C	24	CCG	X	45	AGT	S
4	TTG	D	25	CAT	Y	46	AGC	T
5	TCT	E	26	CAC	Z	47	AGA	U
6	TCC	F	27	CAA	A	48	AGG	V
7	TCA	G	28	CAG	B	49	GTT	W
8	TCG	H	29	CGT	C	50	GTC	X
9	TAT	I	30	CGC	D	51	GTA	Y
10	TAC	J	31	CGA	E	52	GTG	Z
11	TAA	K	32	CGG	F	53	GCT	0
12	TAG	L	33	ATT	G	54	GCC	1
13	TGT	M	34	ATC	H	55	GCA	2
14	TGC	N	35	ATA	I	56	GCG	3
15	TGA	O	36	ATG	J	57	GAT	4
16	TGG	P	37	ACT	K	58	GAC	5
17	CTT	Q	38	ACC	L	59	GAA	6
18	CTC	R	39	ACA	M	60	GAG	7
19	CTA	S	40	ACG	N	61	GGT	8
20	CTG	T	41	AAT	O	62	GGC	9
21	CCT	U	42	AAC	P	63	GGA	.
						64	GGG	SPACE

1.3.1 Encryption Algorithm

Round 1:

- Each letter in the plaintext is converted into its ASCII code.
- Each letter in the plaintext is converted into its ASCII code.
- The binary code will be split into two bits each.
- Each two bits of the binary code will be replaced by its equivalent DNA nucleotides from Table 1.1.

Round 2:

- From the derived DNA strand, three nucleotides will be combined to form a Codon.
- Each Codon will be replaced by its equivalent from the Lookup Table 1.3.

Round 3:

- The derived replaceable characters will be converted into ASCII code.

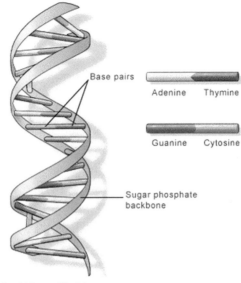

Base pairs

Adenine   Thymine

Guanine   Cytosine

Sugar phosphate backbone

U.S. National Library of Medicine

Fig. 1.1 DNA structure [6]

- Again, the ASCII codes will be converted into its equivalent binary code.
- Again, the binary code will be split into two bits each.
- Each two bits of the binary code will be replaced by its equivalent DNA nucleotide from Table 1.1. The DNA strand so generated will be the final ciphertext (Fig. 1.1).

1.3.2 Decryption Algorithm

The process of reversing the steps from last to first in all rounds continuously will create decryption algorithm.

The algorithm uses the following three levels to complete the encryption. It is briefly described in the Fig. 1.2.

1.4 Algorithm Implementation

1.4.1 Encryption

Let us take the plaintext $M =$ Desire

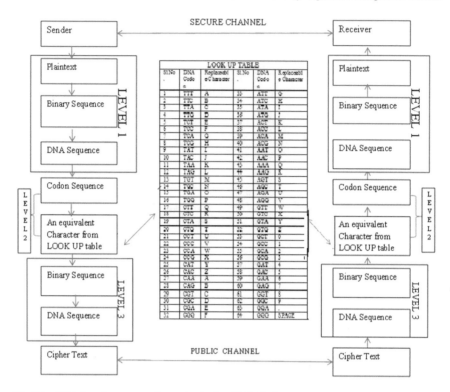

Fig. 1.2 DNA-based cryptography method using DNA Codons

Round 1:

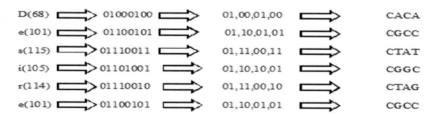

D(68)	⟹	01000100	⟹	01,00,01,00	⟹	CACA
e(101)	⟹	01100101	⟹	01,10,01,01	⟹	CGCC
s(115)	⟹	01110011	⟹	01,11,00,11	⟹	CTAT
i(105)	⟹	01101001	⟹	01,10,10,01	⟹	CGGC
r(114)	⟹	01110010	⟹	01,11,00,10	⟹	CTAG
e(101)	⟹	01100101	⟹	01,10,01,01	⟹	CGCC

Round 2:
The DNA strand from the above round is

$$\text{CACACGCCCTATCGGCCTAGCGCC}$$

Split them into three nucleotides which are called as Codons, and assign equivalent replaceable character from the Lookup table (Table 1.3).

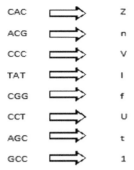

Round 3:

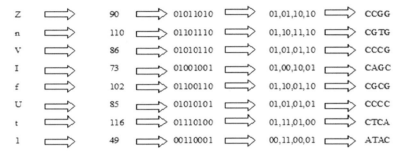

The ciphertext is CCGGCGTGCCCGCAGCCGCGCCCCCTCAATAC

1.4.2 Decryption

The process is done from last to first round to get the plaintext.

In this algorithm three letters forming a Codon and hence all characters the plaintext contains, divisible by 3 is only encrypted into ciphertext. If the characters that are in the plaintext leaves a remainder, when divided by three can be converted with the help of padding to display as ciphertext.

If the remainder is 1, we will pad four zeros at the end when plaintext is transformed into binary data. If the remainder is 2, we will pad two zeros at the end when plaintext is transformed into binary data.

1.5   Experimental Results

1.5.1   Encryption Process

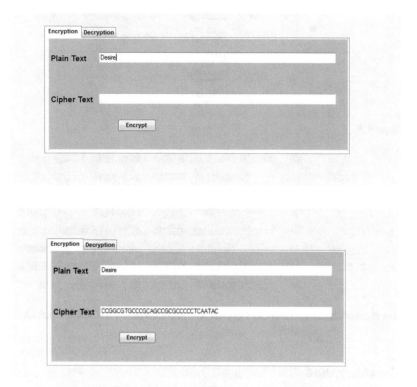

1.5.2   Decryption Process

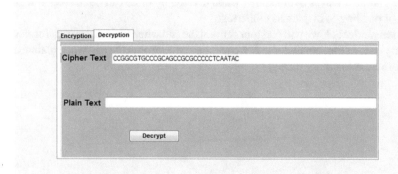

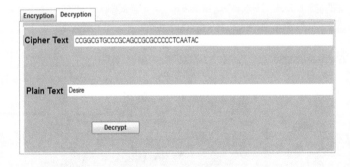

1.5.3 *Padding of Bits*

The encryption and decryption processes are the same as above in all the cases.

1.5.3.1 Encryption Process

If the number of characters of a plaintext is not divisible by 3, then

(1) Pad 4 (zeros) bits when the remainder is 1.
(2) Pad 2 (zeros) bits when the remainder is 2.

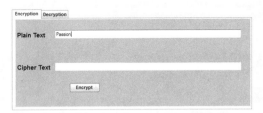

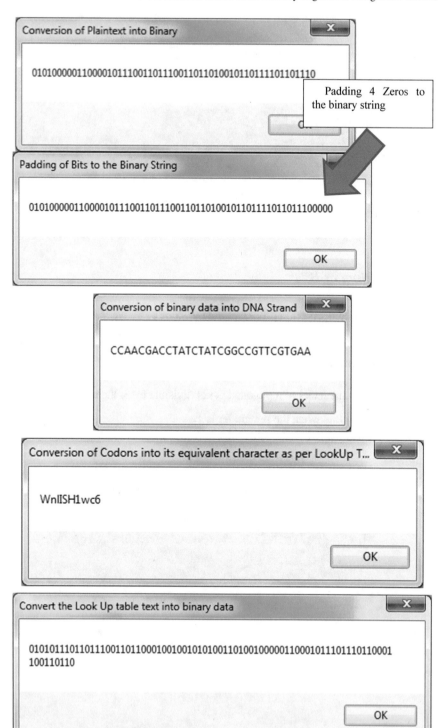

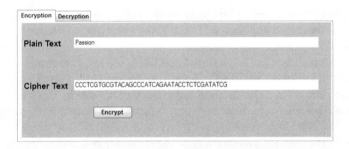

1.5.3.2   Decryption Process

If the number of characters of a plaintext is not divisible by 3, then

(1)   Remove 4 (zeros) bits when the remainder is 1.
(2)   Remove 2 (zeros) bits when the remainder is 2.

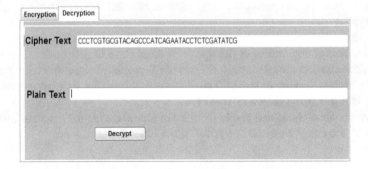

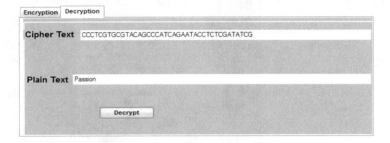

1.6   Result Analysis

Let the sender send the ciphertext in the form of DNA to the receiver end.

Suppose the length of plaintext is "m". Three cases can be discussed here.

Case 1: The plaintext (m) is divisible by 3:

When the plaintext (m) is converted into DNA, the length is increased to $m * 4$, say $m1$. In the second level, the DNA nucleotides are divided into Codons. So, the length is $m1/3$, say $m2$. In the third level, the Codons can be replaced with their equivalent replaceable character from the Lookup table (Table 1.3). Again these can be converted into DNA which is our ciphertext of length $m2 * 4$, say $m3$.

Case 2: The number of characters of plaintext (m) is not divisible by 3, and it leaves the remainder 1:

Then we add additional 2 nucleotides to make a Codon. So, $m1 = m * 4 + 2$ and $m2$, $m3$ is calculated similarly.

Case 3: The number of characters of plaintext is not divisible by 3 and leaves the remainder 2:

We add additional 1 nucleotide to make a Codon. Here, $m1 = m * 4 + 1$ and $m2$, $m3$ is calculated similarly.

Based upon $m1$, $m2$, and $m3$, we calculate the length of ciphertext in each level. The final length of cipher is $m3$. Hence, the time complexity of the encryption process is $O(m)$, and the same process is done in the receiver end also so that the time complexity of decryption process is $O(m)$.

The simulations are performed by using .net programming on Windows 7 system. The hardware configuration of the system used is Core i3 processor/4 GB RAM. The following table shows the performance of the proposed algorithm with different sets of plaintext varying in length. The observations from the simulation have been plotted in Fig. 1.3 and shown in Table 1.4.

From the above table and graph, it can be observed that as the length of plaintext increased, the encryption and decryption times have also increased.

Fig. 1.3 Performance analysis of an algorithm based on length and characters

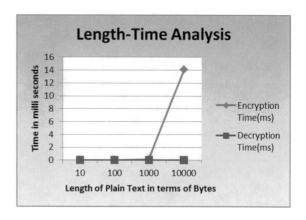

Table 1.4 Length–time analysis

S. No.	Length of plaintext (in terms of bytes)	Encryption time (ms)	Decryption time (ms)
1	10	0.0043409	0.0001647
2	100	0.0123234	0.0006070
3	1000	0.1116644	0.0020742
4	10,000	14.0743528	0.0333596

1.7   Conclusions

Security plays a vital role in transferring the data over different networks, and several algorithms were designed to enhance the security at various levels of network. In the absence of security, we cannot assure the users to route the data freely. In traditional cryptography, as the earlier algorithms could not provide security as desired, we have in full confidence, drawn the present algorithm based on DNA cryptography, that is developed using DNA Codons. It assures a perfect security for plaintext in modern technology as each Codon can be replaced in 64 ways from the Lookup table randomly.

References

1. Adleman LM (1994) Molecular computation of solution to combinatorial problems. Science, New Series, 266(5187):1021–1024
2. Gehani A, Thomas L, Reif J (2004) DNA-based cryptography-in aspects of molecular computing. Springer, Berlin, Heiderlberg, pp 167–188
3. Nixon D (2003) DNA and DNA computing in security practices-is the future in our genes? Global information assurance certification paper, Ver.1.3
4. Popovici C (2010) Aspects of DNA cryptography. Ann Univ Craiova, Math Comput Sci Ser 37(3):147–151
5. Babu ES, Nagaraju C, Krishna Prasad MHM (2015) Light-weighted DNA based hybrid cryptographic mechanism against chosen cipher text attacks. Int J Inf Process 9(2):57–75
6. http://www.chemguide.co.uk/organicprops/aminoacids/doublehelix.gif
7. Yamuna M, Bagmar N (2013) Text Encryption using DNA steganography. Int J Emerg Trends Technol Comput Sci (IJETTCS) 2(2)
8. Reddy RPK, Nagaraju C, Subramanyam N (2014) Text Encryption through level based privacy using DNA Steganography. Int J Emerg Trends Technol Comput Sci (IJETTCS) 3(3):168–172
9. http://www.chemguide.co.uk/organicprops/aminoacids/dnacode.gif
10. Sabry M, Hasheem M, Nazmy T, Khalifa ME (2010) A DNA and amino acids-based implementation of playfair cipher. Int J Comput Sci Inf Secur 8(3)

Chapter 2
Cognitive State Classifiers for Identifying Brain Activities

B. Rakesh, T. Kavitha, K. Lalitha, K. Thejaswi
and Naresh Babu Muppalaneni

Abstract The human brain activities' research is one of the emerging research areas, and it is increasing rapidly from the last decade. This rapid growth is mainly due to the functional magnetic resonance imaging (fMRI). The fMRI is rigorously using in testing the theory about activation location of various brain activities and produces three-dimensional images related to the human subjects. In this paper, we studied about different classification learning methods to the problem of classifying the cognitive state of human subject based on fMRI data observed over single-time interval. The main goal of these approaches is to reveal the information represented in voxels of the neurons and classify them in relevant classes. The trained classifiers to differentiate cognitive state like (1) Does the subject watching is a word describing buildings, people, food (2) Does the subject is reading an ambiguous or non ambiguous sentence and (3) Does the human subject is a sentence or a picture etc. This paper summarizes the different classifiers obtained for above case studies to train classifiers for human brain activities.

Keywords Classification · fMRI · Support vector machines · Naïve Bayes

2.1 Introduction

The main issue in cognitive neuroscience is to find the mental faculties of different tasks, and how these mental states are converted into neural activity of brain [1]. The brain mapping is defined as association of cognitive states that are perceptual with patterns of brain activity. fMRI or ECOG is used to measure persistently with multiunit arrays of brain activities [1]. Non-persistently, EEG and NIRS (Near Infrared Spectroscopy) are used for measuring the brain functions. These development machines are used in conjunction with modern machine learning and pattern recognition techniques for decoding brain information [1]. For both clinical and research purposes, this fMRI technique is most reputed scheme for accessing the brain topography. To find the brain regions, the conventional univariate analysis of fMRI data is used, the multivariate analysis methods decode the stimuli, and cognitive

© The Author(s) 2019   15
Ch. Satyanarayana et al., *Computational Intelligence and Big Data Analytics*,
SpringerBriefs in Forensic and Medical Bioinformatics,
https://doi.org/10.1007/978-981-13-0544-3_2

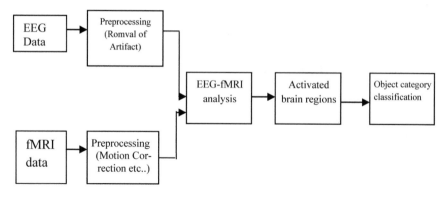

Fig. 2.1 Architecture of fMRI-EEG analysis

states the human from the brain fMRI activation patterns [1]. The multivariate analysis methods use various classifiers such as SVM, naïve Bayes which are used to decode the mental processes of neural activity patterns of human brain. Present-day statistical learning methods are used as powerful tools for analyzing functional brain imaging data.

After the data collection to detect cognitive states, train them with machine learning classifier methods for decoding its states of human activities [2]. If the data is sparse, noisy, and high dimensional, the machine learning classifiers are applied on the above-specified data.

Combined EEG and fMRI data are used to classify the brain activities by using SVM classification algorithm. For data acquisition, EEG equipment, which compatible with 128 channel MR and 3 T Philips MRI scanners, is used [3]. These analyses give EEG-fMRI data which has better classification accuracy compared with fMRI data alone.

2.2 Materials and Methods

2.2.1 fMRI-EEG Analysis

The authors proposed an approach in combination with electroencephalography (EEG) and functional magnetic resonance imaging (fMRI) to classify the brain activities. The authors used support vector machine classification algorithm [4]. The authors used EEG equipment which compatible with 128 channel MR and also 3 T Philips MRI scanners for data acquisition [4]. The analysis showed that the EEG-fMRI data has better classification accuracy than the fMRI data stand-alone (Fig. 2.1).

Based on stimulus property, each voxel regression is performed to identify the signal value. Hidden Markov models like Hojen-Sorensen and Rasmussen are used to analyze fMRI data [1]. These models could not describe the stimulus but they recovered the state as hidden state by HMM. The other way to analyze fMRI data is unsupervised learning.

2.2.2 Classification Algorithms

2.2.2.1 Naïve Bayes Classifier

This classifier is one of the widely used classification algorithms. This is one of the statistical and statistical methods for classification [1]. It predicts the conditional probability of attributes. In this algorithm, the effect of one attribute Xi is independent of other attributes. This is called as conditional independence, and this algorithm is based on Bayes' theorem [1]. To compute the probability of attributes $X_1, X_2, X_3, X_4, ..., X_n$ of a class C this Bayes' theorem is used and it can perform classifications. The posterior probability by Bayes' theorem can be formulated as:

$$P = \frac{\text{likelihood} \times \text{prior}}{\text{evidence}}$$

2.2.2.2 Support Vector Machine

Support vector machines are commonly used for learning tasks, regression, and data classification. The data classifications are divided into two sets, namely training and testing sets. Training set contains the class labels called target value and several observed variables [1–3]. The main goal of this support vector machine is used to find the target values of the test data.

Let us consider the training attributes X_i, where $i = \{1, 2, ... n\}$ and training labels $z\{I, -1\}$. The test data labels can be predicted by the solution of the below given optimization problem

$$\min_{w,p,\in} \frac{1}{2} W^T W + C \sum_{i=1}^{l} \in_i$$

Providing $z_i \left(W^T \phi(X_i) + b \right) \geq 1 - \varepsilon_i$.

Where $\varepsilon_i \geq 0$, ϕ is hyperplane for separating training data, C is the penalty parameter, z_i belongs to $\{1, -1\}$ which is vector of training data labels [4]. The library support vector machines are used for classification purpose, and it solves the support vector machine optimization problems. Mapping of the training vectors X_i into the higher dimensional space can lead to the finding of linear separating hyperplane by the support vector machine [5]. The error term penalty parameter can be represented by $C > 0$.

Table 2.1 Classifiers error rates

Study	Example per class	Selection of features	GNB	SVM	kNN
Sentence versus Picture	40	Yes	0.183	0.112	0.190
	40	No	0.342	0.341	0.381
Categories of semantics	32	Yes	0.081	NA	0.141
	32	No	0.102	NA	0.251
Ambiguity of syntactic	10	Yes	0.251	0.278	0.341
	10	No	0.412	0.382	0.432

2.2.2.3 *K*-Nearest Neighbor Classifier

The k-nearest neighbor classifier is the simplest type of classifier for pattern recognition. Based upon its closest training examples, it classifies the test examples and test example label is calculated by the closest training examples labels [6]. In this classifier, Euclidean distance is used a distance metric.

Let us consider, the Euclidian can be represented as E, and then

$$E^2(b, x) = (b - x)(b - x)'$$

where x and b are row vectors with m features.

2.2.2.4 Gaussian Naïve Bayes Classifier

For fMRI observations, the GNN classifier uses the training data to estimate the probability distribution based on the humans cognitive states [7]. It classifies new example $Y' = \{Y_1, Y_2, Y_3 \ldots Y_n\}$ for the probability estimate $P(c_i|Y')$ of cognitive state c_i for given fMRI observations.

The probability can be estimated by using the Bayes' rule

$$M(c_i|Y') = \frac{M(c_i)\pi_j M(x_j|c_i)}{\sum [M(c_k)\pi_j M(x_j|c_k)]}$$

where M represents estimated distributions by GNB from the training data. All the distributions of the form $M(x_j|c_i)$ is modeled as univariate Gaussian. The variance and mean can be derived from the training data. $M(c_i)$ distributions are modeled as Bernoulli (Table 2.1).

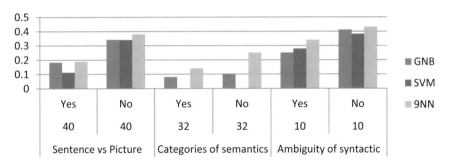

Fig. 2.2 Comparative analysis of GNB, SVM, kNN

2.3 Results

The results shown in the table give better performing variants of GNB, and we can observe the table GNB and SVM classifiers outperformed kNN. The performance generally improves the increasing values of k (Fig. 2.2).

2.4 Conclusion

In this paper, we represented results from three different classifiers of fMRI studies representing the feasibility of training classifiers to distinguish a variety of cognitive states. The comparison indicates that linear support vector machine (SVM) and Gaussian naive Bayes (GNB) classifiers outperform K-nearest neighbor [2]. The accuracy of SVMs increases rapidly than the accurateness of GNB as the dimension of data is decreased through feature selection. We found that the feature selection methods always enhance the classification error in all three studies. For the noisy, high-dimensional, sparse data, feature selection is a significant aspect in the design of classifiers [1]. The results showed that it is possible to use linear support vector machine classification to accurately predict a human observer's ability to recognize a natural scene photograph [8]. Furthermore, classification provides a simply interpretable measure of the significance of the informative brain activations: the quantity of accurate predictions.

References

1. Ahmad RF, Malik AS, Kamel N, Reza F (2015) Object categories specific brain activity classification with simultaneous EEG-fMRI. IEEE, Piscataway
2. Mitchell et al (2004) Learning to decode cognitive states from brain images. Kluwer Academic Publishers, Dordrecht

3. Tom M, Mitchell et al (2008) Predicting human brain activity associated with the meanings of nouns. Science 320:1191. https://doi.org/10.1126/science.1152876
4. Rieger et al (2008) Predicting the recognition of natural scenes from single trial MEG recordings of brain activity.
5. Taghizadeh-Sarabi M, Daliri MR, Niksirat KS (2014) Decoding objects of basic categories from electroencephalographic signals using wavelet transform and support vector machines. Brain topography, pp 1–14
6. Miyapuram KP, Schultz W, Tobler PN (2013) Predicting the imagined contents using brain activation. In: Fourth national conference on computer vision pattern recognition image processing and graphics (NCVPRIPG) 2013, pp 1–3
7. Haxby JV, Gobbini MI, Furey ML, Ishai A, Schouten JL, Pietrini P (2001) Distributad and overlapping representations of faces and objects in ventral temporal cortex. Science 293:2425–2430
8. Cox DD, Savoy RL (2003) Functional magnetic resonance imaging (fMRI) "brain reading": detecting and classifying distributed patterns of fMRI activity in human visual cortex. NeuroImage 19:261–270

Chapter 3
Multiple DG Placement and Sizing in Radial Distribution System Using Genetic Algorithm and Particle Swarm Optimization

M. S. Sujatha, V. Roja and T. Nageswara Prasad

Abstract The present day power distribution network is facing a challenging role to cope up for continuous increasing of load demand. This increasing load demand causes voltage reduction and losses in the distribution network. In current years, the utilization of DG technologies has extremely inflated worldwide as a result of their potential benefits. Optimal sizing and location of DG units near to the load centers provide an effective solution for reducing the system losses and improvement in voltage and reliability. In this paper, the effectiveness of genetic algorithm (GA) and particle swarm optimization (PSO) for optimal placement and sizing of DG in the radial distribution system is discussed. The main advantage of these methods is computational robustness. They provide an optimal solution in terms of improvement of voltage profile, reliability, and also minimization of the losses. They provide the best resolution in terms of improvement of voltage profile, reliability, and also minimization of the losses. The anticipated algorithms are tested on IEEE 33- and 69-bus radial distribution systems using multi-objective function, and results are compared.

Keywords Distributed generation · Genetic algorithm · Particle swarm optimization · Multi objective function · Optimum location · Loss minimization · Voltage profile improvement

3.1 Introduction

The operation of the power system network is too complicated, especially in urban areas, due to the ever-increasing power demand and load density. In the recent past, hydro-, atomic, thermal, and fossil fuel-based generation power plants were in use to meet the energy demands. Centralized control system is used for the operation of such generation systems. Long-distance transmission and distribution systems are used for delivering power to meet the demands of consumers. Due to the depletion of conventional resources and increased transmission and distribution costs, conventional power plants are on the decline [1].

© The Author(s) 2019
Ch. Satyanarayana et al., *Computational Intelligence and Big Data Analytics*,
SpringerBriefs in Forensic and Medical Bioinformatics,
https://doi.org/10.1007/978-981-13-0544-3_3

Distributed generation (DG) is an alternative solution to overcome various power systems problems such as generation, transmission and distribution costs, power loss, voltage regulation [2]. Distributed generation is generation of electric power in small-scale on-site or near to the load center. Several studies revealed that there are potential benefits from DG [3]. Though the DG has several benefits, the most difficulty in placement of DG is that the choice of best location, size, and range of DG units. If the DG units are not properly set and sized, it results in voltage fluctuations, upper system losses, and raise in operational prices. To reduce the losses, the best size and location are very important [4–8]. The selection of objective function is one of the factors that influence the system losses [9–11]. In addition to loss reduction, reliability is also an important parameter in DG placement [12, 13]. In recent years, many researchers have projected analytical approaches based on stability and sensitivity indices to locate the DG in radial distribution systems [14–18]. Viral and Khatod [17] projected a logical method for sitting and sizing of DGs. With this method, convergence can be obtained in a few iterations but is not suitable for unbalanced systems. By considering multiple DGs and different load models, loss reduction and voltage variations are discussed in [19–23]. The authors of [24] considered firefly algorithm for finding DG placement and size to scale back the losses, enhancement of voltage profile, and decrease of generation price. But the drawback of this method is slow rate of convergence. Bat algorithm and multi-objective shuffled bat algorithms are used in [25] and [26], respectively, for optimal placement and sizing of DG in order to meet the multi-objective function. Mishra [27] proposed DG models for optimal location of DG for decrease of loss. Optimization using genetic algorithm is proposed in [28]. The particle swarm optimization is presented in [29] for loss reduction. The authors of [30] discussed the minimization of losses by optimal DG allocation in the distribution system.

This paper is intended to rise above all drawbacks by considering multi-objective function for most favorable sizing and placing of multi-DG with genetic algorithm and particle swarm optimization techniques, and the effectiveness of these algorithms is tested on different test systems.

3.2 DG Technologies

The power-generating systems located close to the consumer primacies and linked directly to the distribution network are called as DG. Distributed generation resources (DGRs) can be classified into renewable DG resources and conventional DG resources. Examples for the renewable DG resources are solar and wind turbines and those for the conventional DGs are combustion engines, reciprocating engines, etc. [22].

3.2.1 Number of DG Units

The placing of DGs may be either single or multiple DG units. Here, multiple DG approach is considered consisting of three DGs in the test system of radial distribution system for loss reduction [22].

3.2.2 Types of DG Units

Based on power-delivering capacity, the DGs are divided into four types [22].
   Type-1: DG injecting active power at unity PF. Ex. Photovoltaic; Type-2: DG injecting reactive power at zero PF. Ex. gas turbines; Type-3: DG injecting active power but consuming reactive power at PF ranging between 0 and 1. Ex. wind farms; Type-4: DG injecting both active and reactive powers at PF ranging between 0 and 1. Ex. Synchronous generators.

3.3 Mathematical Analysis

3.3.1 Types of Loads

Loads may be of different types, viz., domestic loads, commercial loads, industrial loads, and municipal loads.

3.3.2 Load Models

In the realistic system, loads do not seem to be only industrial, commercial, and residential; but it is depending on nature of the area being supplied [21]. Practical voltage-depended load models are considered in [8]. The recently developed numerical equations for different load models in the system are specified by Eqs. (1) and (2).

$$P_{Di} = P_{DO}\left(P_1\left(\frac{V_i}{V_0}\right)^\alpha + q_1\left(\frac{V_i}{V_0}\right)^\alpha r_1\left(\frac{V_i}{V_0}\right)^\alpha + s_1\left(\frac{V_i}{V_0}\right)^\alpha \right) \tag{1}$$

$$Q_{Di} = Q_{D0}\left(P_2\left(\frac{V_i}{V_0}\right)^\beta + q_2\left(\frac{V_i}{V_0}\right)^\beta r_2\left(\frac{V_i}{V_0}\right)^\beta + s_2\left(\frac{V_i}{V_0}\right)^\beta \right) \tag{2}$$

where
   P_{Di} is active power demand, Q_{Di} is reactive power demand at Bus I, P_{D0i} and Q_{D0i} at Bus I are demand operating points of active and reactive powers, V_0 is voltage at the operating point, V_i is Bus I voltage, and β and α show the reactive and active

Table 3.1 Load models of exponent values

Type of load	Exponents		Load type	Exponents	
Constant load	$\alpha_0 = 0$	$\beta_0 = 0$	Residential load	$\alpha_r = 0.92$ 4.04	$\beta_r =$
Industrial load	$\alpha_i = 0.18$	$\beta_i = 6$	Commercial load	$\alpha_c = 1.51$ 3.4	$\beta_c =$

power exponents for industrial, commercial, residential, and constant load models with subscriptions i, c, r, and 0, respectively.

Table 3.1 indicates exponent values of the different load models. p_1, q_1, r_1, s_1 are the weight coefficients of active power and p_2, q_2, r_2, s_2 are the weight coefficients of reactive power respectively. The exponent values of coefficients for different loads and types are specified as below:

Type-1: Constant load: $p_1 = 1, q_1 = 0, r_1 = 0, s_1 = 0, p2 = 1, q_2 = 0. r_2 = 0, s_2 = 0$.
Type-2: Industrial load: $p_1 = 0, q_1 = 1, r_1 = 0, s_1 = 0, p2 = 0, q_2 = 1. r_2 = 0, s_2 = 0$.
Type-3: Residential load: $p_1 = 0, q_1 = 0, r_1 = 1, s_1 = 0, p2 = 0, q_2 = 0. r_2 = 1, s_2 = 0$.
Type-4: Commercial load: $p_1 = 0, q_1 = 0, r_1 = 0, s_1 = 1, p2 = 0, q_2 = 0. r_2 = 0, s_2 = 1$.
Load type-5: Mixed or practical load: $p_1 = $ta1, $q_1 = $ta2, $r_1=$ta3, $s_1 = $ta4 and $P_2 = $trl, $q_2 = $tr2, $r_2 = $tr3, $s_2 = $tr4. Also for the practical mixed load models, ta1 + ta2 + ta3 + ta4 = 1 and tr 1 + tr2 + tr3 + tr4 = 1.

3.3.3 Multi-objective Function (MOF)

The multi-objective function given in (3) can be used for most favorable position and sizing of multi-DG using GA and PSO techniques.

$$MOF = C_1 PLI + C_2 QLI + C_3 VDI + C_4 RI + C_5 SFI \tag{3}$$

where PLI, QLI, VDI, RI, and SFI are active power loss, reactive power loss, voltage deviation, reliability, and sensitivity or shift factor of the system, explained in Eqs. (8)–(12), respectively. $C_1, C_2, C_3, C_4,$ and C_5 are the weight factors of the indices of the system. The active and reactive losses are given in the following Eqs. (4) and (5).

- Real power loss (PL)

$$PL = \sum_{K=1}^{N} |I_K|^2 R_K \tag{4}$$

• Reactive power loss (*QL*)

$$QL = \sum_{K=1}^{N} |I_K|^2 X_K \qquad (5)$$

• The ENS can be obtained as [19]

$$ENS = \alpha d \sum_{K=1}^{N} \lambda_k |I_{KP}| V_{\text{rated}} \qquad (6)$$

where I_k is branch current, R_k resistance; X_k reactance of *k*th line [27], I_{KP} peak load branch current, λ_k failure rate for *k*th branch or line, V_{rated} system rated voltage, α load factor, and *d* repair duration.

• The reliability of the system is given as [12]

$$R = 1 - \left(\frac{ENS}{PD}\right) \qquad (7)$$

where *R* is reliability and *PD* total power demand.

3.3.4 Evaluation of Performance Indices Can Be Given by the Following Equations

$$PLI = \frac{PL_{DG}}{PL_{NO-DG}} \qquad (8)$$

$$QLI = \frac{QL_{DG}}{QL_{NO-DG}} \qquad (9)$$

$$VDI = \max_{j=2}^{N} \left(\frac{V_{\text{reff}} - V_{DGj}}{V_{\text{reff}}}\right) \qquad (10)$$

where *N* is total number of buses, V_{reff} reference voltage, and V_{DGj} DG system voltage.

The equation of SFI by fitting of DG of suitable size can be expressed as [13]

$$SFI = \max_{\substack{j-1 \\ j \neq \text{slackbus} \\ \& \; PV\text{bus}}}^{x-1} \left|\frac{\Delta x_j}{x_{injj}}\right| \qquad (11)$$

The equation of *RI* is given as [19]

$$RI = \frac{ENS_{DG}}{ENS_{NO-DG}} \tag{12}$$

where ENS_{DG} is ENS by DG and ENS_{No-DG} is ENS by without DG condition.

3.4   Proposed Methods

In this work, GA and PSO are the evolutionary techniques used for the planning of DGs.

3.4.1   Genetic Algorithm (GA)

It is one of the evolutionary algorithm techniques [10]. It is a robust optimization technique based on natural selection. In case of DG placement, fitness function can be loss minimization, voltage profile improvement, and cost reduction.

Figure 3.1 shows the flow sheet to seek out optimum sitting and size of DGs using GA in different test systems, where PDG, QDG, and LDG represent real and reactive powers and position of DG and are considered in the form of population.

3.4.2   Particle Swarm Optimization (PSO)

PSO is a population-based calculation and was presented by Dr. Kennedy and Dr. Eberhart in 1995. It is a biologically inspired algorithm [15]. It is a novel intelligence search algorithm [9] that provides good solution to the nonlinear complex optimization problem. Figure 3.2 demonstrates the procedure flowchart to discover ideal sitting and estimating of DGs in the different test frameworks utilizing PSO.

3.5   Results and Discussions

Results of best placing and size of multi-DG with MOF using GA and PSO in a particular type of system are presented in the following sections. MATLAB (2011a) is used to validate the proposed methodologies.

3.5.1   33-Bus Radial Distribution System

A 33-bus radial distribution system connection diagram is shown in Fig. 3.3 The aggregate active and reactive power loads of standard IEEE 33-bus test framework are 3.72 MW and 2.30 MVAr [18]. The results for 33-bus radial distribution system are specified in Tables 3.2, 3.3, 3.4 and 3.5 and Figs. 3.4, 3.5 and 3.6.

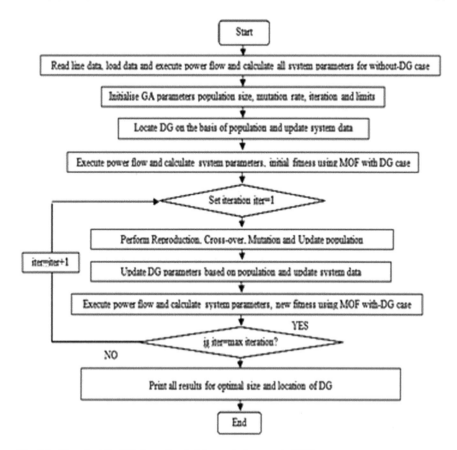

Fig. 3.1 Flowchart for GA for optimal sizing and placement of DG

Table 3.4 shows that all type-4 DGs are suitable and are ideally introduced in 33-bus radial distribution system by utilizing GA and PSO techniques with various load models.

The active and reactive power loss reductions of 33-bus multi-DG structure are 89.45 and 87.58% for GA and 92.76 and 91.15% for PSO, respectively, and they are shown in Table 3.6. Voltage profiles of 33-bus radial distribution system for various load models are shown in Fig. 3.5. By observing this, the voltage profiles are better with DG than without DG and are farther improved with DG-PSO compared with DG-GA and without DG.

From Fig. 3.4 and Table 3.5, it is observed that the losses are very much less for DG-PSO compared with no DG and DG-GA for different load models. Figure 3.6 shows that reliability is improved and ENS is decreased with DG-GA for different load models of 33-bus radial distribution system. Hence, the proposed methodologies are more effective for reducing losses and ENS and improving the reliability of all load models.

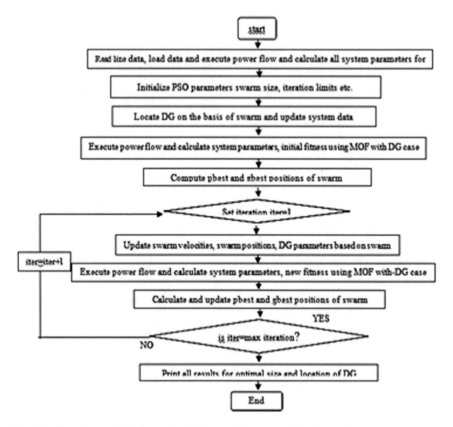

Fig. 3.2 Flowchart for PSO for optimal sizing and placement of DG

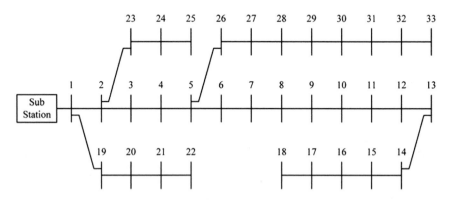

Fig. 3.3 Line diagram of 33-bus radial distribution system

Table 3.2 ENS and reliability of 33-bus radial distribution system with multi-DG using GA and PSO

Load type	Parameters	No DG	DG-GA	DG-PSO
Constant	ENS (MW)	0.1243	0.0103	0.0102
	Reliability	0.9658	0.9970	0.9972
Industrial	ENS (MW)	0.1038	0.0102	0.0038
	Reliability	0.9715	0.9991	0.9986
Residential	ENS (MW)	0.1068	0.0178	0.0035
	Reliability	0.9650	0.9945	0.9987
Commercial	ENS (MW)	0.1068	0.0091	0.0048
	Reliability	0.9691	0.9972	0.9983
Mixed	ENS (MW)	0.1055	0.0102	0.0032
	Reliability	0.9705	0.9995	0.9991

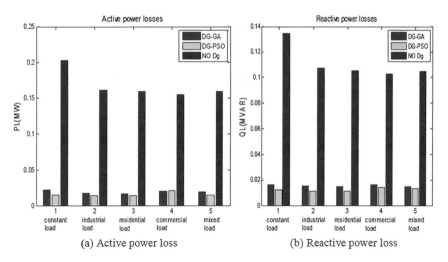

(a) Active power loss   (b) Reactive power loss

Fig. 3.4 Active and reactive power losses with multi-DG and different load models of 33-bus radial system

3.5.2   69-Bus Radial Distribution System

Figure 3.7 shows the 69-bus radial distribution system. The total active and reactive power loads of the system are 3.802 MW and 2.695 MVAr, respectively. The data for the system is taken from [14, 18]. Tables 3.6, 3.7, 3.8 and 3.9 and Figs. 3.8 and 3.9 show the results of the system using GA and PSO for multi-DG and mixed load models. From Table 3.7, it is observed that all DGs are type-4 at different optimal locations.

Table 3.3 **MOF** and indices for 33-bus radial distribution system with multi-DG and different loads

Load type	Fitness (MOF)	PLI	QLI	VDI	RI	SFI	Optimal technology
Constant	0.1861	0.1051	0.1215	0.0125	0.0817	1.0549	GA
	0.1635	0.0765	0.0905	0.0115	0.0812	1.0067	PSO
Industrial	0.1995	0.1190	0.1449	0.0115	0.0958	1.0411	GA
	0.1718	0.0887	0.1106	0.0071	0.0385	1.0525	PSO
Residential	0.1989	0.1109	0.1359	0.0123	0.0172	1.0021	GA
	0.1755	0.0971	0.1215	0.0056	0.0349	1.0322	PSO
Commercial	0.2039	0.1280	0.1551	0.0121	0.0875	1.0301	GA
	0.1951	0.1210	0.1603	0.0063	0.0137	1.0210	PSO
Mixed	0.1978	0.1196	0.1435	0.0118	0.069	1.0311	GA
	0.1736	0.0949	0.1155	0.0055	0.0306	1.0321	PSO

Table 3.4 Size and location of multiple DG in 33-bus radial distribution system

Load type	DG1		DG2		DG3		Location	DG type	Optimal technology
	P(MW)	Q (Mvar)	P(MW)	Q (Mvar)	P(MW)	Q (Mvar)			
Constant	1.4813	0.9862	0.9451	1.0961	0.4311	1.0961	3,13,30	All type-4	GA
	0.9460	1.0073	0.5660	0.2148	0.5862	0.2148	30,15,25	All type-4	PSO
Industrial	1.3504	0.9014	0.7791	0.9252	0.6479	0.5281	3,30,14	All type-4	GA
	1.1855	0.6591	0.9301	0.8462	0.8102	0.5731	24,30,14	All type-4	PSO
Residential	1.7453	0.9859	0.9702	1.7432	0.6938	0.3641	3,30,14	All type-4	GA
	1.2466	0.8898	0.6271	0.5972	1.0705	0.5813	30,14,24	All type-4	PSO
Commercial	1.953	1.4071	1.0778	1.0072	0.7401	0.4301	3,30,14	All type-4	GA
	0.6709	0.3705	1.1479	0.9129	1.3751	1.3658	14,30,24	All type-4	PSO
Mixed	1.0285	1.0508	1.5631	1.1938	0.7408	0.4102	30,3,14	All type-4	GA
	0.7052	0.3953	0.9843	1.2085	1.5203	0.4493	14,30,24	All type-4	PSO

Table 3.5 Active and reactive power losses of 33-bus radial distribution system with and without DG

Load type	Losses	No DC	DG-GA	DC-PSO
Constant	PL (MW)	0.2019	0.0213	0.0146
	QL (MVAR)	0.1345	0.0167	0.0119
	Loss reduction			
	% PL	–	89.45	92.76
	% QL	–	87.58	91.15
Industrial	PL (MW)	0.1611	0.0181	0.0138
	QL (MVAR)	0.1075	0.0152	0.0112
	Loss reduction			
	%PL	–	88.76	91.43
	%QL	–	85.86	89.58
Residential	PL (MW)	0.1589	0.0168	0.0133
	QL (MVAR)	0.1054	0.0149	0.0132
	Loss reduction			
	%PL	–	89.01	91.30
	%QL	–	85.23	87.77
Commercial	PL (MW)	0.1552	0.0201	0.0146
	QL (MVAR)	0.1031	0.0162	0.0119
	Loss reduction			
	%PL	–	87.74	86.59
	%QL	–	84.28	86.42
Mixed	PL (MW)	0.1593	0.0195	0.0147
	QL (MVAR)	0.1057	0.0151	0.0132
	Loss reduction			
	%PL	–	87.75	90.77
	%QL	–	85.71	87.51

Table 3.6 MOF and indices for 69-bus radial distribution system with multi-DG

Load type	Fitness (MOF)	PLI	QLI	VDI	RI	SFI	Optimal technology
Mixed load	0.1612	0.0689	0.1168	0.0148	0.0015	1.0478	GA
	0.1486	0.0522	0.1075	0.0123	0.0015	1.0001	PSO

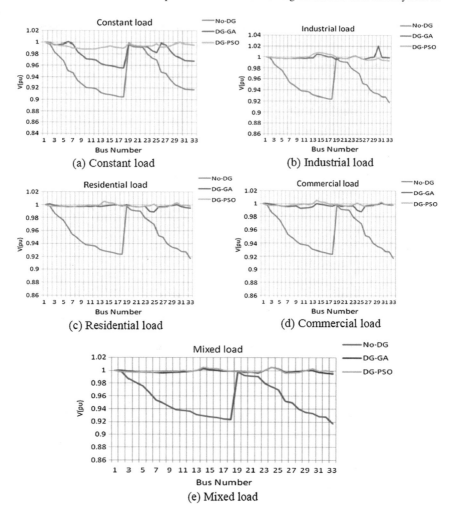

Fig. 3.5 Voltage profiles of 33-bus radial system for different load models

The reduction of losses is 93.18 and 88.19% for GA and 95.00 and 89.46% for PSO, respectively, which are shown in Table 3.8. The results reveal that the GA and PSO are more effective methods for loss reduction in radial distribution systems. It can also be noted that ENS is reduced and reliability is improved by placing of multi-DG in 69-bus radial distribution system using GA and PSO, which are given in Table 3.9. Figures 3.8 and 3.9 show that the active and reactive power losses are reduced and voltage profile is improved in 69-bus system using proposed methods.

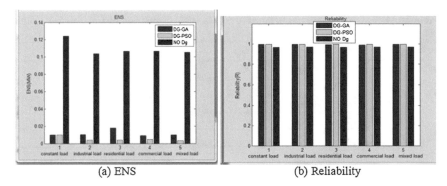

(a) ENS   (b) Reliability

Fig. 3.6 Energy not supplied and reliability of 33-bus radial distribution system

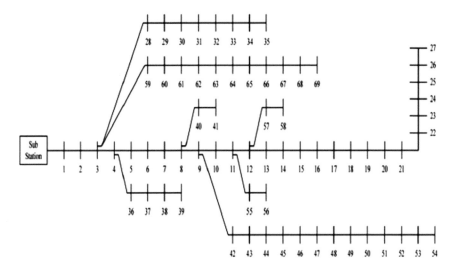

Fig. 3.7 A 69-bus radial distribution system

Table 3.7 Size and location for 69-bus radial distribution system with multi-DG

Load type	DG1		DG2		DG3		Location	DG type	Optimal technology
	P(MW)	Q(Mvar)	P(MW)	Q(Mvar)	P(MW)	Q(Mvar)			
Mixed load	2.3030	2.4463	1.2769	0.8510	1.6290	1.2199	4,11,61	All type-4	GA
	3.3861	2.6942	0.7239	0.5564	1.7114	1.0616	4,12,61	All type-4	PSO

Table 3.8 Active and reactive power losses of 69-bus radial distribution system with and without DG

Load type	Losses	No DG	DG-GA	DG-PSO
Mixed load	PL (MW) QL	0.1703	0.0116	0.0085
	(MVAR)	0.0788	0.0093	0.0083
	Loss reduction			
	% PL	–	93.18	95.00
	% QL	–	88.19	89.46

Table 3.9 ENS and reliability of 69-bus radial distribution system with multi-DG

Load type	Parameters	No DG	DG-GA	DG-PSO
Mixed load	ENS (MW) reliability	0.2337	0.0003	0.0003
		0.9367	0.9996	0.9998

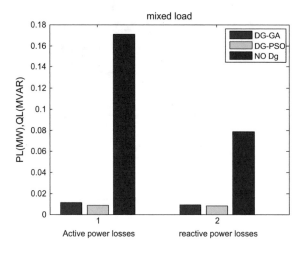

Fig. 3.8 Active and reactive power losses of 69-bus radial distribution system

3.6   Conclusions

In this paper, genetic algorithm and particle swarm optimization strategies are executed to find the ideal size and location of the DG with multi-objective function
based on the indices. The proposed strategies are applied to 33- and 69-bus radial
distribution systems for different types of loads and multi-DGs. The results show
that the projected procedure is more effective for power loss reduction, improvement
of voltage profile, and system reliability.

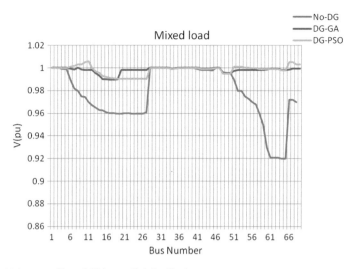

Fig. 3.9 Voltage profiles of 69-bus radial distribution system

References

1. Bohre AK, Agnihotri G (2016) Optimal sizing and sitting of DG with load models using soft computing techniques in practical distribution system. IET Gener Transm Distrib pp. 1–16
2. Ackermann T, Andersson G, Soder L (2001) Distributed generation: a definition. Electr Power Syst Res 57:195–204
3. Chidareja P (2004) An approach to quantify technical benefits of distributed generation. IEEE Trans Energy 19(4):764–773
4. Viral R, Khatod DK (2012) optimal planning of DG systems in distribution system: a review. Int J Renew Sustain Energy Rev 16:5146–5165
5. Kalambe S, Agnihotri G (2014) Loss minimization techniques used in DN bibliographic survey. Int J Renew Sustain Energy Rev 29:184–200
6. Georgilakis Pavlos S, Hatziargyriou Nikos D (2013) Optimal distributed generation placement in power distribution networks: models, methods, and future research. IEEE Trans Power Syst 28(3):3420–3428
7. Ghosh S, Ghosh S (2010) Optimal sizing and placement of DG in a network system. Electr Power Energy Syst 32:849–856
8. Ghosh N, Sharma S, Bhattacharjee S (2012) A load flow based approach for optimal allocation of distributed generation units in the distributed network for voltage improvement and loss minimization. Int J Comput Appl 50(15):0975–8887
9. Singh D, Verma KS (2009) Multi objective optimization for DG planning with load models. IEEE Trans Power Syst 24(1):427–436
10. Ochoa LF, Padilha-Feltrin A, Harrison GP (2006) Evaluating DG impacts with a multi objective index. IEEE Trans Power Deliv 21(3):1452–1458
11. Pushpanjalli M, Sujatha MS (2015) A Navel multi objective under frequency load shedding in a micro grid using genetic algorithem. Int J Adv Res Electr Electron Instrum Eng 4(6) ISSN:2278–8875
12. Al-Muhaini M, Heydt GT (2013) Evaluating future power distribution system reliability including distributed generation. IEEE Trans Power Deliv 28(4):2264–2272

13. Chowdhury AA, Agarwal SK, Koval DO (2003) Reliability modeling of distributed generation in conventional distribution systems planning and analysis. IEEE Trans Ind Appl 39(5):1493–1498
14. Samui A, Singh S, Ghose T et al (2012) A direct approach to optimal feeder routing for radial distribution system. IEEE Trans Power Deliv 27(1):253–260
15. Acharya N, Mahat P, Mithulananthan N (2006) An analytical approach for DG allocation in primary distribution network. Int J Electric Power Energy Syst 28(10):669–678
16. Gopiya Naik SN, Khatod DK, Sharma MP (2015) Analytical approach for optimal siting and sizing of distributed generation in radial distribution networks. IET Gener Transm Distrib 9(3):209–220
17. Viral R, Khatod DK (2015) An analytical approach for sizing and siting of DGs in balanced radial distribution networks for loss minimization. Electric Power Energy Syst 67:191–201
18. Nahman JM, Dragoslav MP (2008) Optimal planning of radial distribution networks by simulated annealing technique. IEEE Trans Power Syst 23(2):790–795
19. Hien CN, Mithulananthan N, Bansal RC (2013) Location and sizing of distributed generation units for load ability enhancement in primary feeder. IEEE Syst J 7(4):797–806
20. Hung DQ, Mithulananthan N (2013) Multiple distributed generator placement in primary distribution networks for loss reduction. IEEE Trans Ind Electron 60(4):1700–1708
21. El-Zonkoly AM (2011) Optimal placement of multi-distributed generation units including different load models using particle swarm optimization. Swarm Evol Comput 1(1):50–59
22. Kansal S, Vishal K, Barjeev T (2013) Optimal placement of different type of DG sources in distribution networks. Int J Electric Power Eng Syst 53:752–760
23. Zhu D, Broadwater RP, Tam KS et al (2006) Impact of DG placement on reliability and efficiency with time-varying loads. IEEE Trans Power Syst 21(1):419–427
24. Nadhir K, Chabane D, Tarek B (2013) Distributed generation and location and size determination to reduce power losses of a distribution feeder by Firefly Algorithm. Int J Adv Sci Technol 56:61–72
25. Behera SR, Dash SP, Panigrahi BK (2015) Optimal placement and sizing of DGs in radial distribution system (RDS) using Bat algorithm. In: International conference on circuit, power and computing technologies (ICCPCT), 2015, 19–20 March, 2015, pp 1–8, https://doi.org/10.1109/iccpct.2015.7159295
26. Yammani C, Maheswarapu S, Matam SK (2016) A multi-objective Shuffled Bat algorithm for optimal placement and sizing of multi DGs with different load models. Electr Power Energy Syst 79:120–131
27. Mishra M (2015) Optimal placement of DG for loss reduction considering DG models. IEEE International conference on electrical, computer and communication technologies (ICECCT), 2015, 5–7, pp 1–6 (March 2015)
28. Haupt RL, Haupt SE Practical genetic algorithms, 2nd edn. Wiley, inc, Hoboken, New Jersey, Published simultaneously in Canada (2004)
29. Bohre AK, Agnihotri G, Dubey M et al (2014) A novel method to find optimal solution based on modified butterfly particle swarm optimization. Int J Soft Comput Math Control (IJSCMC) 3(4):1–14
30. Roja V, Sujatha MS (2016) A review of optimal dg allocation in distribution system for loss minimization. J Electr. Electron Eng e-ISSN: 2278–1676, p-ISSN: 2320-3331, vol 2, pp. 15–22

Chapter 4
Neighborhood Algorithm for Product Recommendation

P. Dhana Lakshmi, K. Ramani and B. Eswara Reddy

Abstract Web mining is the process of extracting information directly from the Web by using data mining techniques and algorithms. The goal of Web mining is to find the patterns in Web data by collecting and analyzing the information. Most of the business organizations are interested to recommend the products in a quick and easy manner. The main goal of recommender system of a product is to increase the rate of customer retention and enhance revenue of an organization. In this paper, neighborhood algorithm-based recommender system is developed by considering both product and seller ratings with an objective to reduce time for getting review of an appropriate product.

Keywords Web mining · Product recommendation · Neighborhood principle
Product rating · Seller rating

4.1 Introduction

E-commerce is a commercial transaction which is conducted electronically on the Internet. E-commerce has shown enormous growth in the last few years because it is very convenient to shop anytime, anywhere, and in any device. Searching the products online is much simpler and more efficient than searching the products in stores. Customers use the E-commerce to gather the information and to purchase or sell the goods in a trouble-free manner. In order to increase the sales of an E-commerce organization and to attract more number of customers, a filtering system has been introduced which is known as recommender system.

E-commerce organization needs a business house which is used to manufacture and maintain the products, an online Web service to provide a good turn of products to the customer and the customer to buy and sell the products. Recommender systems filter the vital information out of a large amount of dynamically generated information according to user's preferences, interest, or observed behavior about item. Recommender system also predicts and improves the decision making process based on user preferred item. Recommender systems also improve the decision-making pro-

© The Author(s) 2019
Ch. Satyanarayana et al., *Computational Intelligence and Big Data Analytics*,
SpringerBriefs in Forensic and Medical Bioinformatics,
https://doi.org/10.1007/978-981-13-0544-3_4

cess which is beneficial to both service providers and users. Recommender system was defined from the perspective of E-commerce as a tool that helps the user to search through available products which are related to users' interest and preference. The relevant user's information is stored in the Web, and the extraction of knowledge from the Web is used to build a recommender system. The entire process of extracting the information from the Web is known as Web mining.

Recommender systems use three different types of filtering methods: content-based filtering, collaborative filtering, and hybrid filtering.

Content-based technique is a domain-dependent algorithm. Using this technique, the most positively rated documents such as Web pages and news are suggested to the user [1]. The recommendation is made based on the features extracted from the content of the items the user has evaluated in the past. CBF uses vector space model, probabilistic models, decision trees, and neural networks for finding the similarity between documents in order to generate meaningful recommendations [2, 3]. Collaborative filtering builds a database which consists of all user selected preferences as a separate list. It is a domain-independent prediction method. The recommendations are then made by calculating the similarities between the users and build a group named as neighborhood [4, 5]. User gets recommendations to those items that he has not rated before but that were already positively rated by users in his neighborhood. Collaborative filtering (CF) technique [6] may be classified as memory-based and model-based.

Memory-based CF can be achieved in two ways: user-based and item-based techniques. For a single item, user-based collaborative filtering method suggests the items by comparing the ratings given by the users. In item-based filtering techniques instead of calculating similarity between users, it considers the similarity between weighted averages of items.

Model-based CF is used to learn the model to improve the performance of the collaborative filtering technique. It will recommend the products quickly which are similar to neighborhood-based recommendation by applying machine algorithms [7].

To optimize the recommender system, a technique called hybrid filtering technique is developed which combines different recommendations for suggesting the efficient products to the customer [8].

This paper is organized as follows: Sects. 4.2 and 4.3 discusses about related work, existing system respectively. Section 4.4 introduces a novel integration technique for recommending products, Sects. 4.5 and 4.6 gives experimental results and conclusion.

4.2 Related Work

Recommendations in E-commerce are made by considering a parameter such as cost, color, rating. Earlier, the recommendations were made based on single user preferences, which have led to less accurate results. There are three approaches used in recommendation systems: content, collaborative, and hybrid filtering techniques.

Collaborative filtering technique recommends the items to the target user based on considering the other user opinion with similar taste. Unlike collaborative, content-based filtering technique recommends the products by considering single user's information. Hybrid filtering is the combination of two or more filtering techniques which is used to increase the accuracy and performance of recommender systems. A new recommendation method [5] collects the preferences on individual items from all users and suggests the items to the target user based on calculating the similarity between items by clustering methods [6]. This technique is not suitable for real-time applications. To overcome this problem, Top-n new items are recommended to the customer by developing a user specific feature based similarity model. Similarity between users is calculated using cosine measure. This method is not applicable to actual ratings given by the user in a Web site; it considers only binary preferences on items. The solution for this is to understand the customers repeat purchase items for recommending the products to the customers [1, 9]. This technique mainly focused on the customer support and repeating product purchase of the customer. This technique is very difficult to calculate utilitarian and hedonic values numerically, and it also requires profile and history of last visited customers to a Web site [3].

To predict the new products for a customer a maximum entropy and grey decision making methods are developed [10]. The main principle behind this algorithm is trustworthiness. Trustworthiness of the recommender system refers to the degree by which system conforms to people's expectations by supporting the software life cycle and by providing services in each stage of the life cycle [11]. The main limitation of this method is inherent in complexity of objects. [5, 12, 13] developed a new product recommendation technique by considering public opinion using graph theory and sentiment analysis. This sentiment analysis is performed only by using fixed length, and also, it cannot be suitable to variable length in real-time applications. No one technique is developed to display optimum no. of products to the Web users. Therefore, novel techniques to achieve optimum no. of products by considering both product and seller rating with minimum time are very much necessary.

4.3 Existing System

Due to the increase in online shopping, the normal customer shops more times at an online store before the store can profit from that customer. Based on the customer support and repeating the product purchase of the customer the success of the store depends. Utilitarian value and hedonic value are the hypotheses to affect repeat purchase intention positively. The main drawback of this method is computation of these two hypothesis completely depends on profile of the customer. The recommendation approach using collaborative filtering method is carried out by analyzing the structural relationships among three parties (P, O, X) and classifying the relationships into eight categories in which four are stable and four are unstable which are shown in Fig. 4.1.

Case1: O is a friend of P; X is a friend of O; therefore X is a friend of P with large probability.

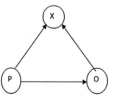

Case 2: O is enemy of P; X is enemy of O; therefore X is friend of P.

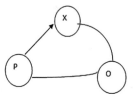

Case 3: O is friend of P; X is enemy of O; therefore X is enemy of P.

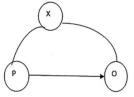

Case 4: O is enemy of P; X is friend of O; therefore X is enemy of P

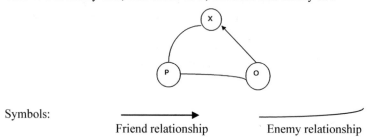

Symbols:

Friend relationship   Enemy relationship

Fig. 4.1 Structural relationship model for users

Considering the above four cases, the similarity between different users and based on the threshold values, the products are recommended to the target user. The similarity is computed based on the product rating given by the user. The existing system does not address the cold start problem, loss of neighbor transitivity and sparsity.

4.4 Proposed System

To increase the sales and to attract the customer, it is necessary to improve the performance of the recommender system. Drawbacks present in the collaborative filtering method are solved by using the neighborhood algorithm by considering the product rating as well as the seller rating. The product rating is given by the customer or the end user who buys that product. Rating is given based on the quality of the product and the satisfaction of the end user on the product. By using the product rating, the customer can easily judge whether the product is good or bad.

The seller rating is given a group of people. They include the admin of the E-commerce Web site, various sellers, manufacturer of the product, and many more. The seller rating is given based on the packing, delivery, and quality of the product. By using the seller rating, the admin can monitor his sellers, and at the same time, it will help the customer to understand the shipping details and quality in a very clear manner. Seller rating is computed by calculating the average of all the seller ratings, user ratings, and manufacturer ratings. When only product rating is considered for recommending the products, the following problems have occurred:

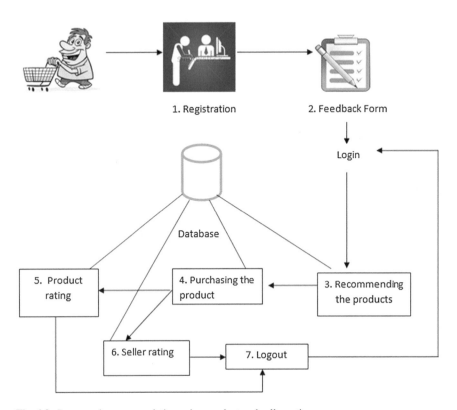

Fig. 4.2 Integrated recommendation using product and seller rating

1. The cold-start problem
2. Loss of neighbor transitivity
3. Sparsity

To solve these three problems, the neighborhood algorithm with integrated rating approach has been introduced. The cold-start problem can be avoided by including a feedback form in the application where the user can enter his preferences. Based on the feedback, the products are recommended to the user when he enters into the application for the first time.

A. **System Design**

The integrated recommendation using product and seller rating overcomes the problem of cold-start problem, data scarcity problem, synonymy, and single attribute consideration. The process of the application is elaborated in Fig. 4.2.

B. **Algorithm**

INPUTS:

USER {$user_1$, $user_2$... $user_n$}: a set of users present in the application system.
PRODUCT {P_1, P_2... P_n}: a set of products present in the database.
PRODUCT_PURCHASE {PP_1, PP_2... PP_n}: a set of purchased products by $user_i$.
FEEDBACK {F_1, F_2.... F_n}: a set of feedback forms for each $user_i$ in USER.
P_RATING {$user_1$, $user_2$... $user_n$}: set of product ratings given by user.
S_RATING{$user_1$, $user_2$... $user_n$}: set of seller ratings given by user.
FRIEND{$user_1$, $user_2$... $user_n$}: set of similar users for $user_{target}$.
ENEMY{$user_1$, $user_2$... $user_n$}: set of dissimilar users for $user_{target}$.
$user_{target}$: target user to whom the products are recommended.
Similarity $_{product}$: similarity between two users when only product rating is considered.
Similarity $_{seller}$: similarity between two users when only seller rating is considered.

OUTPUT:

REC_ITEM: set of product items recommended to $user_{target}$.
PROCEDURE:
BEGIN
If user_login = first_time.
Call feedback_recommendation
Else
Call integrated_recommendation.
End If.
Method: Feedback_recommendation.
1: For each $user_i \in$ USER
       1.1: For each $F_j \in$ FEEDBACK($user_i$)
Display products present in F_j.
       End for.
   End for.

2: Move user$_i$ product ratings into set P_RATING(user$_i$).
3. Move user$_i$ seller ratings into set S_RATING(user$_i$).

Method: Integrated_recommendation.
1: set the threshold value for enemy is p.
2: For each user$_i$ € USER .
2.1: calculate the similarity between the user$_{target}$ and user$_i$.
Sim (usertarget, user$_i$)
2.2: If Sim (user$_{target}$, user$_i$) $\leq$ p
      put user$_i$ into set ENEMY(user$_{target}$)
Else
3: Move user$_i$ seller ratings into set S_RATING(user$_i$).
      put user$_i$ into set FRIEND(user$_{target}$).
End if.
End for.
For each user$_i$ $\in$ ENEMY(user$_{target}$) do
3.1: For each user$_j$ $\in$ USER do
3.1.1: calculate Sim (user$_i$, user$_j$)
3.2: If Sim (user$_i$, user$_j$) $\geq$ P then
Move user$_j$ into set ENEMY(user$_{target}$)
Else
Movet user$_j$ into set FRIEND(user$_{target}$)
End if.
   End for.
End for.

The algorithm first allows the user to register in the system, and upload a feedback form for several attributes if the user is login at first time. Then, it checks whether the user login is for the first time or not. If it is for the first time, then the algorithm invokes the feedback_recommendation method. If the user login is not for the first time, then the integrated_recommendation method is called.

The feedback_recommendation method displays the products to the customers based on the feedback form. The integrated_recommendation method first sets a threshold value p or enemy. USER$_i$ is the user set, friend is the similar users set, and enemy is the dissimilar users set present in the system. Then, for the target user similarity is calculated for each user present in the system. If the similarity computed is less than the threshold value, then the user is kept in the enemy set of the target user; otherwise, the user is kept in the friend set. After classifying the users into different sets, the same procedure is repeated by considering the seller rating. Finally, the mean of both the similarities is calculated, and if the similarity is greater than the 60%, the products are recommended to the target user.

1. Feedback-based recommendation

The feedback form is introduced for every customer after the registration. This is used when the customer logins for the first time and when there are no similar products found for a particular customer.

2. Integration recommendation using neighborhood algorithm.

2.1. Calculating similarity using product rating

Determine user target's "possible friends" through "enemy's enemy is a friend" and "enemy's friend is an enemy" rules. The similarity between the target user and normal user is computed by using the product rating.

2.2 Calculating similarity using seller rating

Determine user target's "possible friends" through "enemy's enemy is a friend" and "enemy's friend is an enemy" rules. The similarity between the target user and normal user is computed by using the seller rating.

3. Integration and recommendation

The mean value of similarities calculated in the above two modules is calculated, and if the mean value is greater than 60%, then the algorithm recommends the user$_i$ products to the target user.

Example

Let us consider an n*m matrix for each customer C_i which consists of the product IDs purchased by C_i, his/her product and seller ratings on the purchased products.

$$C_i = \begin{bmatrix} P_1 & PR_1 & SR_1 \\ P_2 & PR_2 & SR_2 \\ P_3 & PR_3 & SR_3 \\ . & . & . \\ . & . & . \\ P_m & PR_n & SR_n \end{bmatrix}$$

In the above matrix, the C_i represents the customer IDs, $P_{1,2...m}$ represents the product IDs, PR_i denotes the product ratings, and SR_i denotes seller ratings where $i \in \{1, 2. 3,...n\}$.

Let us consider there are total five existing customers present in the system, and let C_{tar} be the target user. The product ID set is $\{p1, p1, ... p6\}$ = {DELL, HP, CELKON, APPLE, NOKIA, LENOVO}. The n*m matrix for the six customers is given as below:

$$C_1 = \begin{bmatrix} P_1 & 4 & 3 \\ P_2 & 3 & 5 \\ P_3 & 3 & 3 \end{bmatrix}$$

$$C_2 = \begin{bmatrix} P_2 & 2 & 4 \\ P_4 & 3 & 4 \end{bmatrix}$$

$$C_3 = \begin{bmatrix} P_1 & 2 & 4 \\ P_3 & 1 & 4 \\ P_4 & 4 & 4 \end{bmatrix}$$

$$C_4 = \begin{bmatrix} P_3 \ 5 \ 5 \end{bmatrix}$$

$$C_{tar} = \begin{bmatrix} P_1 \ 4 \ 5 \\ P_2 \ 5 \ 4 \end{bmatrix}$$

Similarity Formula

$$\text{sim}\left(C_{target}, C_i\right) = \frac{\sum_{\text{pro\_item}\in I}\left(R_{\text{target } i} - \overline{R_{\text{target}}}\right) * \left(R_{ij} - \overline{R_i}\right)}{\sqrt{\sum_{\text{pro\_item}j\in I\text{target}}\left(R_{\text{target}-j} - \overline{R_{\text{target}}}\right)^2} * \sqrt{\sum_{\text{pro\_item}j\in I i}\left(R_{i-j} - \overline{R_i}\right)^2}}$$

$$(4.1)$$

Notations

- set I represents the common product items that were rated by C_{target} and C_i where C denotes the customer.
- I_{target} and I_i denote the product item sets that were rated by user$_{target}$ and user$_i$.
- $R_{target-j}$ and R_{i-j} represent user$_{target}$'s ratings and C_i's ratings on product item$_{pro\_item}$.
- R_{target} and R_i denote the average rating scores of all the product items rated by C_{target} and C_i.

1. Calculate the similarity between C_{target} and C_1 using Eq. (4.1).

$$\text{Sim}(C_{tar}, C_1) = \frac{(4 - 4.5) * (4 - 3.6) + (5 - 4.5) * (3 - 3.6)}{\sqrt{(4 - 4.5)^2 + (5 - 4.5)^2} * \sqrt{(4 - 3.6)^2 + (3 - 3.6)^2 + (4 - 3.6)^2}} = \frac{0.5}{0.7} = 0.71$$

2. Calculate the similarity between the users using the seller rating using Eq. (4.1).

$$\text{Sim}(C_{tar}, C_1) = \frac{(5 - 4.5) * (5 - 4.6) + (4 - 4.5) * (5 - 3.6)}{\sqrt{(5 - 4.5)^2 + (4 - 4.5)^2} * \sqrt{(5 - 4.6)^2 + (5 - 4.6)^2 + (4 - 4.6)^2}} = \frac{0.4}{0.5} = 0.8$$

Now, the mean value of both the similarities calculated through step 1 and 2 is:

$$\text{Mean} = \frac{(0.71 + 0.8)}{2} = 0.78$$

Therefore, the similarity between the C_{tar} and C_1 is 78%. Since it is greater than 60%, C_1 is considered as friend of C_{tar} and product P_3 is recommended to C_{tar}.

C. **Physical Model**

The neighborhood relationship categorizes the customers into friends and enemies for a target user. The classification is done based on the similarity values computed between the target user and the existing user. The relationship is described in Fig. 4.3.

In Fig. 4.4, an example of the user and product item level in the integrated recommendation is described. At the user level, the customers already existing and the target

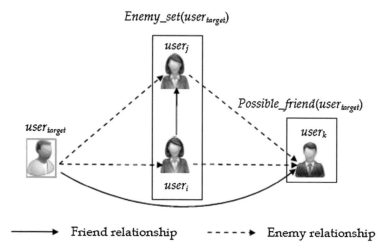

Fig. 4.3 Neighborhood relationship between customers

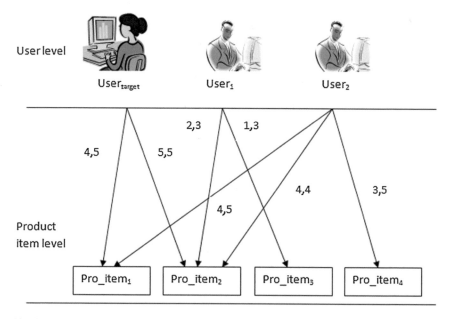

Fig. 4.4 User and product level for integrated recommendation

user are present. In the product item level, the products purchased by the customers are present. The relationship from the user level to product item level consists of the product and seller rating given by the particular user for his/her purchased products.

The similarity between the $user_{target}$ and $user_2$ is high, so the product pro_$item_4$ is recommended to the $user_{target}$. The $user_2$ is now friend of $user_{target}$. The similarity between the $user_{target}$ and $user_1$ is very less, and hence, he is considered as an enemy, and $user_2$ purchased products are not recommended to the $user_{target}$.

Fig. 4.5 Registration page

4.5 Experiments and Results

The whole process of Web recommendation using integration rating is for predicting online purchase behavior of user which is displayed in this paper. In order to describe the online purchase behaviors comprehensively, six features are extracted from the seven fields of the raw data with SQL Server.

In Fig. 4.5, the application allows the customer to register his details which is done only once. The details of the customer are stored in the database when the user clicks on registration button.

In Fig. 4.6, feedback form consisting of four fields, brand, color, prize, and range, is given to the customer. The feedback form appears after the registration page. The main purpose of including the feedback form is to recommend the products to the customer when he enters into the application for the first time. The entire data is stored in the database when the customer clicks on upload feedback button.

In Fig. 4.7, the login page for the customer has been created. The login page will start a session for the customer, using which he/she can purchase the products.

In Fig. 4.8, the products present in the database are displayed based on the feedback form. The feedback-based recommendation is used when the user enters the application for the first time. The customer can give the product rating in this page. This reduces the problem of cold-start.

In Fig. 4.9, if the customer wants to buy the product, then after clicking on buy now button, the customer will directly connect to the Flipkart Web site where he can buy the product.

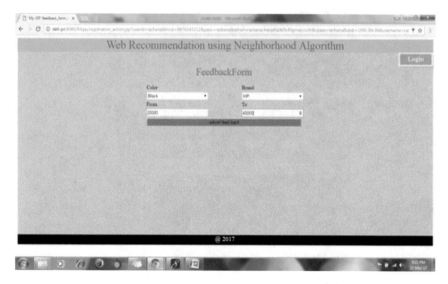

Fig. 4.6 Feedback form

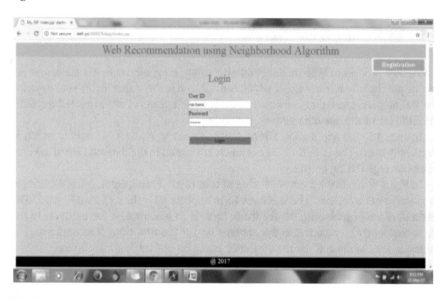

Fig. 4.7 Login page

In Fig. 4.10, the list of purchased products for a particular customer is displayed in this page. The customer here can give the seller rating based on some factors, such as packing, delivery.

When the customer logins for the second time, the items are recommended based on the ratings. In Fig. 4.11, the product recommendation is done based on the product

Fig. 4.8 Recommendation based on feedback form

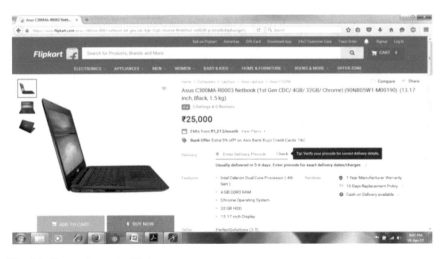

Fig. 4.9 Connecting to the Flipkart

rating. And in the integrated recommendation, the products are displayed based on both the product and seller rating. The more accurate results are shown in integrated recommendation than in product recommendation.

The results above are given with an assumption that every user has purchased only one product. Table 4.1 consists of the total no. of products recommended when only product rating is considered. Similarly, Table 4.2 consists of the total no. of products recommended when both the product and seller rating are used. The results depict a clear picture of the improved efficiency in recommending the product. Figure 4.12 signifies the Web recommendation using product rating. When a single product rating

Fig. 4.10 Giving the seller rating

Fig. 4.11 User second-time login

is used, more number of products is recommended to the user with less accuracy that the customer will like that product. Graph 2 signifies the Web recommendation using the integrated rating. The integrated rating is the combination of product and seller rating. In Fig. 4.13, the less number of products are recommended to the customer with higher accuracy.

Table 4.1 No. of recommended products using product rating

No. of users	No. of products recommended
0	0
100	36
200	65
300	92
400	117
500	149
600	188
700	218
800	242
900	286
1000	328

Table 4.2 No. of recommended products using integrated rating

No. of users	No. of products recommended
0	0
100	25
200	49
300	82
400	101
500	124
600	148
700	172
800	198
900	230
1000	258

Fig. 4.12 Web recommendation using product rating

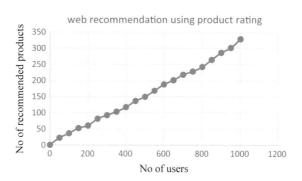

4.6   Conclusion and Future Work

The neighborhood algorithm for product recommendation provides an easy access to the customers to purchase a product in the Web application. The algorithm uses inte-

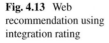

Fig. 4.13 Web recommendation using integration rating

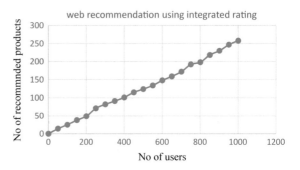

grated recommendation approach where the product and seller rating are considered as two parameters for calculating the similarity between the target user and the existing user. The recommended products are accurate when the integrated approach is implemented with selected feature. In future work, we will develop an automated similarity threshold setting method to satisfy the requirements of different E-commerce users.

References

1. Thotharat N (2014) Thai local product recommendation using ontological content based filtering. Business Comput Dept Fac Manage Sci
2. Liu X, Li J (2014) Using support vector machine for online purchase predication. IEEE Trans
3. Trinh G (2013) Predicting future purchases with the poisson lognormal model. IEEE Trans
4. Jamalzehi S, Menhaj MB (2016) A New similarity measure based on item proximity and closeness for collaborative filtering recommendation. IEEE, 28
5. Cai Y, Leung H, Li Q et al (2015) Typicality-based collaborative filtering recommendation. IEEE Trans
6. Zhong Y, Fan Y, Huang K, Tan W, Zhang J (2014) Time-aware service recommendation for mashup creation. IEEE Trans Serv Comput 8(3):356–368
7. Zheng Z, Ma H, Lyu MR, King I (2013) QoS-aware web service recommendation by collaborative filtering. IEEE Trans
8. Kavinkumar V, Reddy RR, Balasubramanian R, Sridhar M, Sridharan K, Venkataraman D (2013) A hybrid approach for recommendation system with added feedback component. IEEE
9. Chiu C, Wang E, Fang Y, Huang H (2014) Understanding customers' repeat purchase intentions in B2C E—commerce: the roles of utilitarian value, Hedonic Value and Perceived Risk. Inf Syst J 24(1):85–114
10. Tian Y, Ye Z, Yan Y, Sun M (2015) A practical model to predict the repeat purchasing pattern of consumers in the C2C e-commerce. Electron Commerce Res 15(4):571–583
11. Jiang R (2014) A Trustworthiness evaluation method for software architectures based on the principle of maximum entropy (POME) and the grey decision-making method (GDMM). Entropy 16(9):4818–4838

12. Lin S, Lai C, Wu C, Lo C (2014) A trustworthy QoSbased collaborative filtering approach for web service discovery. J Syst Software 93:217–228
13. Leung CH, Li Q et al (2014) Typicality-based collaborative filtering recommendation. IEEE Trans

Chapter 5
A Quantitative Analysis of Histogram Equalization-Based Methods on Fundus Images for Diabetic Retinopathy Detection

K. G. Suma and V. Saravana Kumar

Abstract Contrast enhancement is an essential step to improve the medical images for analysis and for better visual perception of diseases. Fundus images help in screening and diagnosing the diabetic retinopathy. These images must be enhanced in contrast, and the brightness should be preserved to view the features correctly. In addition, the fundus image should be analyzed by Histogram Equalization-based methods to detect DR abnormalities effectively. In this paper, various image enhancements based on Histogram Equalization have been reviewed on fundus images and the results are compared using image quality measurement tools such as absolute mean brightness error to assess brightness preservation, peak signal-to-noise ratio to evaluate the contrast enhancement, entropy to measure richness of the details of the image. The proposed results show that the Adaptive Histogram Equalization is the best enhancement method which helps in the detection of diabetic retinopathy in fundus images.

Keywords Diabetic retinopathy · Fundus images · Histogram Equalization Normalization · Brightness preservation · Contrast enhancement · Entropy · Image quality measurement

5.1 Introduction

Image enhancement is a preprocessing step in image or video processing that is mainly used to increase the low contrast of an image [1]. Contrast enhancement adjusts dark or bright pixels of an image to bring out the hidden feature in that image. Digital medical images help the professional graders in screening and diagnosing the diseases. Diabetic retinopathy (DR) is a complication of diabetes mellitus that affects the vision of the patient and sometimes leads to blindness. Fundus image captured by fundus camera helps to analyze the anatomy of retinal part of an eye (blood vessels, macula, fovea, optic disk) and is used to monitor the abnormalities of diabetic retinopathy (microaneurysms, hemorrhages, soft and hard exudates, neovascularizations) [2].

© The Author(s) 2019
Ch. Satyanarayana et al., *Computational Intelligence and Big Data Analytics*,
SpringerBriefs in Forensic and Medical Bioinformatics,
https://doi.org/10.1007/978-981-13-0544-3_5

Artifacts in fundus images are often a hurdle to detect the abnormalities. Charge-coupled device in fundus camera might be a cause for the noise similar to digital images [3]. The electrical impulses of photoreceptors present in the retina travel toward the brain, so the optic disk looks brighter than the other retinal regions. This bright nature of the optic disk impedes the detection of bright pixel DR abnormality within the optic disk. In addition, reflex contraction of the iris to the flash from the fundus camera leads to blurred image [3].

The contrast between the blood vessels present in fundus image and the retinal background is very low. Hence, the analysis of tiny retinal vasculature and retinal-related abnormalities is difficult [4]. Therefore, the enhancement of retinal region in fundus image is important which provides better visualization of blood vessels and increases the accuracy to detect the abnormalities. Contrast enhancement-based methods have been investigated in several papers in the past decades. Fadzil et al. [4] studied fluorescein angiogram image instead of fundus image and enhanced the contrast based on the retinal pigments using independent component analysis; the paper concludes that Contrast-Limited-based Adaptive Histogram Equalization is beneficial for vessel-based segmentation. Choukikar et al. [5] proposed a method of enhancing the fundus image using Histogram Equalization with cumulative density function (CDF) and performing threshold-based segmentation for blood vessel extraction. Rahim et al. [6] investigated Histogram Equalization, Contrast-Limited Adaptive Histogram Equalization, and Mahalanobis distance for fundus image. They recommended Mahalanobis distance as the best algorithm for blood vessel enhancement. Green plane of the fundus image presents the dark region with highest contrast against the background of the image. The extraction of green channel of fundus image is used as preprocessing step to detect the DR abnormalities [7, 8].

Improving the contrast of fundus image with brightness preservation might lead to better visualization of smaller components and hidden features in the fundus image. This paper discusses and reviews the use of Bi-Histogram Equalization-based methods, Multi-Histogram Equalization-based methods, and Clipped Histogram Equalization-based methods in fundus imaging and compared them using image quality measurement tool entropy.

5.1.1 Extracting the Fundus Image From Its Background

Fundus images are red in color with black background. The black background does not have any details on fundus and can be removed. Equalizing the images with black background increases the darkness within the details of the image. Therefore, we planned to eliminate the black background as preprocessing. The background black has a pixel value of 0, and the realistic regions have other values. Pixel threshold was fixed such that the region with black background would convert into 1 and other regions would convert into 0. After thresholding, the regions with 0 are replaced with gray scale of input image, and the grayscale pixels are replaced with input fundus image. Therefore, the fundus of an eye alone was available and then the green plane

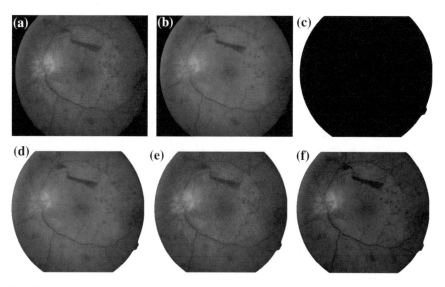

Fig. 5.1 **a** Original fundus image, **b** gray scale of original fundus image, **c** binary image of grayscale image, **d** replace the black pixels with grayscale image pixels, **e** replace the gray pixels with input fundus image pixels, **f** extract the green plane from the extracted color fundus image

of the fundus image was extracted for further enhancement process. The sequence of preprocessing the fundus image is orderly shown in Fig. 5.1.

5.1.2   Image Enhancement Using Histogram Equalization-Based Methods

Histogram Equalization is a contrast enhancement technique that levels and distributes the dynamic range of intensity values uniformly among all histogram bins [10]. The main disadvantage of HE is that the global content of the image gets enhanced. Therefore, HE highlights borders and edges but reduces the local details of the image. Another disadvantage is that it produces over-enhancement and saturation artifacts throughout the image [9].

Conventional Histogram Equalization (CHE) method distributes an equal number of gray levels using cumulative probability density (CPD) equal to 0.5 [1]. CHE produces an output image with a uniformly distributed flattened histogram using CPD. The disadvantage of CHE is that the equalization does not have any differentiation between the various pixels.

5.1.2.1　Bi-Histogram Equalization-Based Methods

Bi-Histogram Equalization methods are presented to overcome the drawbacks of HE. Kim et al. [11] in 1997 proposed brightness preserving Bi-Histogram Equalization method (BBHE) which divides the image histogram as X_r into two sub-images based on the mean intensity value of the input image, and then HE is applied for both the sub-images from X_0 to X_r and X_{r+1} to X_{L-1} independently. BBHE preserves the brightness of image extent better than HE.

Similar to BBHE, in 1999 Wang et al. [12] proposed Dualistic Sub-Image Histogram Equalization (DSIHE) method that divides the image histogram into two sub-images based on the median intensity value of the input, and then HE is applied for both the sub-images independently. BBHE and DSIHE are not much suitable for higher degree of brightness preservation to avoid grating artifacts.

Chen and Ramli, in 2003 [13] proposed Minimum Mean Brightness Error Bi-Histogram Equalization performs threshold level which returns minimum of Absolute Mean Brightness Error. However, the method fails to control the over-enhancement of the image when the images require more brightness preservation. Sengee et al. [14] in 2010 proposed Bi-Histogram Equalization with Neighborhood Metric (BHENM) that separates the large histogram bins into sub-bins using neighborhood metrics to washout artifacts. In 2012, Zuo et al. [15] proposed Range-Limited Bi-Histogram Equalization (RLBHE) that divides the histogram into two by a threshold and minimizes the intra-class variance [15]. Many Bi-HE methods were proposed, which enhance the contrast meaningfully and domain the brightness into some extent, but sometimes the method introduces undesirable artifacts.

5.1.2.2　Multi-histogram Equalization-Based Methods

Multi-Histogram Equalization methods were introduced to enhance the contrast and preserve brightness of an image effectively than HE and Bi-HE. Multi-HE divides the image histogram recursively into several sub-images and then equalizes the sub-histograms independently to get an effective equalization. Chen et al. in 2003 [13] proposed Recursive Mean-Separate Histogram Equalization method as an improvement of BBHE. They separated each new histogram recursively based on their respective means and then equalized all the separated histograms from X_0 to X_{rl}, X_{rl+1} to X_r, X_{r+1} to X_{ru}, X_{ru+1} to X_{L-1} independently.

In 2007, Sim et al. [16] proposed Recursive Sub-Image Histogram Equalization method as an improvement of DSIHE. They separated each new histogram recursively based on their respective median using equal area property with Shannon's entropy to divide the image equally into sub-images and then equalized all the separated histograms independently. Wadud et al. [17] in 2007 proposed Dynamic Histogram Equalization method that eliminates the higher histogram components with domination, toward the lower histogram in the image. This method had shown a smooth enhancement of the image.

Nimkar et al. in 2013 [18] proposed Multi-Decomposition Histogram Equalization method that decomposed the input image into a number of smaller images according to the image application using a unique logic, and then equalization is applied for each sub-image. Finally, the separated and equalized images are interpolated in correct order. This method generates a pleasant look to the final image than the other HE methods. Multi-HE prevents the introduction of undesirable artifacts by preserving the brightness but does not show significant improvement in contrast enhancement of an image.

5.1.2.3 Clipped Histogram Equalization-Based Methods

The HE methods stretch the contrast of high histogram regions without changing the contrast of low histogram regions. While the interested region of an image occupies a small portion, the intensities of the region are pushed toward right or left side of the histogram and will not be successfully enhanced by HE [9]. Clipped Histogram Equalization methods adjust the figure of the input histogram by increasing or reducing the value in the histogram's bins based on the clipping limit (i.e., plateau limit), and then, equalization takes place within each histogram, which restricts the over-enhancement.

Adaptive Histogram Equalization was introduced in order to enhance small region of interest in an image. The method computes several histograms each consistent to a distinct section of the image, by redistributing the lightness values [18]. However, this method amplifies noise in some similar regions of an image. Contrast-Limited Adaptive Histogram Equalization method establishes clipping limit, and the cropped portion will be reordered back to the histogram, and then HE is carried out [18].

5.2 Image Quality Measurement Tools (IQM)—Entropy

Entropy is used to measure the richness of the particulars in the enhanced image Y_k [19]. Entropy is computed as follows for the PDF p,

$$\text{Entropy}[p] = -\sum_{k=0}^{L-1} p(Y_k) \log_2 p(Y_k)$$

5.3 Results and Discussions

Histogram Equalization, Bi-Histogram Equalization-based method (i.e., BBHE), Multi-Histogram Equalization-based methods (i.e., RMSHE, DHE, MDHE), and Clipped Histogram Equalization-based method (i.e., AHE) are compared using IQM

Table 5.1 Entropy: to measure the richness of the details in image

Images	Methods					
	HE	BBHE	Multi-HE-based methods			CHE-based method
			RMSHE	DHE	MDHE	AHE
Image006	5.2668	5.3643	5.7812	5.9478	2.5497	6.533
Image009	5.2723	4.1124	5.4006	5.5447	2.7007	6.2816
Image033	4.5795	4.5932	4.7253	4.8459	2.9999	5.6214
Image034	4.1817	3.9838	4.0878	3.9074	2.3498	4.8998
Image036	4.9749	5.2081	5.2461	5.412	2.6265	6.15
Image041	5.376	5.1101	5.6642	5.8038	2.6804	6.4043
Image042	5.3162	5.5577	5.9655	6.1145	2.7252	6.4315
Image046	5.1016	5.0007	5.2648	5.453	2.7726	6.317
Image047	5.1642	5.1619	5.2959	5.4174	2.6966	6.3208
Image050	5.126	5.2314	5.2686	5.4348	2.6834	6.1671

tool entropy as tabulated in Table 5.1. The test fundus images were from Diaretdb0, Diaretdb1, Messidor online databases and from two eye clinics images. The enhanced DR fundus images (Image046, Image042) of Diaretdb0 database are shown in Fig. 5.2.

HE images provide over-enhancement in highlighted region and reduce the local details of the image (Fig. 5.2b). BBHE images show equalization based on the mean gray-level value; it preserves brightness but does not enhance the details presented in fundus image (Fig. 5.2c). RMSHE images show the details presented within the image to some extent (Fig. 5.2d). DHE smooths the fundus image and does not enhance the details presented in the fundus image (Fig. 5.2e). MDHE images require more enhancement works to get the image to be enhanced (Fig. 5.2f). AHE performs equalization by clipping and distributing the lightness value of an image to all histogram bins (Fig. 5.2g). Therefore, AHE increases the details presented in the fundus image against the background that has been analyzed by entropy measure.

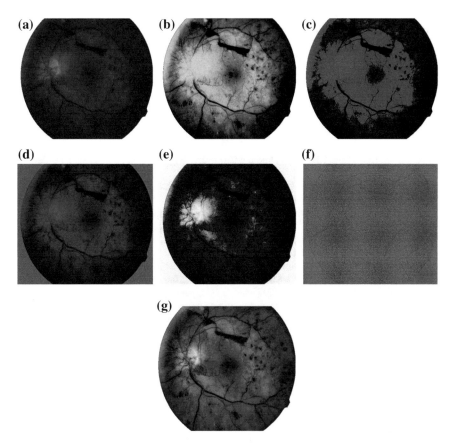

Fig. 5.2 **a** Green plane of extracted fundus image, **b** Histogram Equalization, **c** Brightness preserving Bi-Histogram Equalization, **d** Recursive Mean-Separate Histogram Equalization, **e** Dynamic Histogram Equalization, **f** Multi-Decomposition Histogram Equalization, **g** Adaptive Histogram Equalization

5.4 Conclusion

Enhancing the details present in the fundus image would help us to detect abnormalities of diabetic retinopathy in the retinal part of the eye. Abnormalities of diabetic retinopathy are presented in both dark and bright pixels. Microaneurysms, hemorrhages, and neovascularization are the abnormalities in dark pixels, while exudates and cotton wool spots are the abnormalities in bright pixels. In AHE method, brightness is clipped and equally distributed to all the histogram bins, so it fails to preserve brightness presented in the image. However, the hidden details presented in the fundus image are exposed against the fundus background clearly, which would help us to detect the abnormalities presented in the diabetic retinopathy.

Acknowledgements The authors would like to thank Dr. M. Pratap, Ophthalmologist, Vasan Eye Care, Nagercoil, India, and Dr. Bejan Singh, Ophthalmologist, Bejan Singh Eye Hospital, Nagercoil, India, for help in collecting the images.

References

1. Gonzalez RC, Woods RE (2002) Digital image processing, 2nd edn. Prentice Hall, Englewood Cliffs, NJ
2. Saine PJ, Tyler ME Fundus photography overview. Opthalmic Photogr—Angiography and Electronic Imaging, 2nd edn, Butter Heinemann Medical, ISBN: 0750673729, 2011–2013, Available Online: http://www.opsweb.org/?page=fundusphotography
3. Wong D (1979) Fundus photography and fluorescein angiography, photography of the external eye, photographic artifacts and the patient. J Ophthalmic Photogr 2(1):37–45
4. Fadzil N, Nugroho I Contrast Enhancement of retinal vasculature in digital fundus image. In: International conference on digital image processing. IEEE, New York, pp 137–141
5. Choukikar P, Patel AK, Mishra RS (2014) Segmenting the Optic disc in retinal images using thresholding. Int J Comput Appl (0975-8887) 94(11):6–10
6. Rahim I, Zaki H (2014) Methods to enhance digital fundus image for diabetic retinopathy detection. In: IEEE 10th international colloquium on signal processing & its application (CSPA), March 2014, pp 221–224
7. Goatman KA, Fleming AD, Philip Sam, Williams GJ, Olson JA, Sharp PF (2011) Detection of new vessels on the optic disc using retinal photographs. IEEE Trans Med Imaging 30(4):972–979
8. Hassan SSA, Bong DBL, Premsenthil M (2012) Detection of neovascularization in diabetic retinopathy. J Digit Imaging. Springer, Published online: 7 September 2011, Society for Imaging Informatics in Medicine 2011, Issue 25, pp 436–444
9. Raju A, Dwarakish GS, Reddy DV (2013) A comparative analysis of histogram equalization based techniques for contrast enhancement and brightness preserving. Int J Sign Process Image Process Pattern Recogn 6(5):353–366
10. Umbaugh SE (1998) Computer vision and image processing. Prentice Hall, Englewood Cliffs, NJ, p 209
11. Kim YT (1997) Contrast enhancement using brightness preserving bi-histogram equalization. IEEE Trans Consum Electron 43(1):1–8
12. Wang Y, Chen Q, Zhang B (1999) Image enhancement based on equal area dualistic sub-image histogram equalization method. IEEE Trans Consum Electron 45(1):68–75
13. Chen SD, Ramli A (2003) Contrast enhancement using recursive mean-separate histogram equalization for scalable brightness preservation. IEEE Trans Consum Electron 49(4):1301–1309
14. Sengee N, Sengee A, Choi HK (2010) Image contrast enhancement using bi-histogram equalization with neighborhood metrics. IEEE Trans Consum Electron 56(4):2727–2734
15. Zuo C, Chen Q, Sui X (2012) Range limited bi-histogram equalization for image contrast enhancement. Optik
16. Sim KS, Tso CP, Tan YY (2007) Recursive sub-image histogram equalization applied to gray scale images. Pattern Recogn Lett 28:1209–1221
17. Wadud MAA, Kabir MH, Dewan MAA, Chae O (2007) a dynamic histogram equalization for image contrast enhancement. IEEE Trans Consum Electron 53(2):593–600

18. Nimkar S, Shrivastava S, Varghese S (2013) Contrast enhancement and brightness preservation using multi-decomposition histogram equalization. Signal Image Process: Int J (SPIJ) 4(3):83–93
19. Patel O, Maravi YPS, Sharma S (2013) A comparative study of histogram equalization based image enhancement techniques for brightness preservation and contrast enhancement. Signal Image Process.: Int J (SIPIJ) 4(5):11–25

Chapter 6
Nanoinformatics: Predicting Toxicity Using Computational Modeling

Bhavna Saini and Sumit Srivastava

Abstract The various potential properties of nanomaterials make them prominent for most of the applications, used in our day-to-day life. Due to increase in the use of the nanomaterials, it is anticipated that the manufacturing and use of engineered nanoparticles (NP) will also grow rapidly. These engineered nanoparticles became toxic sometimes during the process of manufacturing as well as utilizing. Hence, there is a vital requirement for early recognition of toxic nature of these nanoparticles. Computational modelling is an effective approach for the same, which predicts toxicity on the basis of previous experimental data or molecular properties. QSAR is an emerging process to envision the toxicity based on their molecular structure properties. The various data mining techniques can accelerate this computation task and help in accurate and fast toxicity prediction. This paper leads the readers to work in this area, by providing a groundwork of the data, tools, and techniques used, along with future research directions.

Keywords Data mining · Nanotoxicity · Nanoinformatics · Computational modelling

6.1 Introduction

Nanotoxicity is the part of nanoscience which works to find out the extent to which the specific nanoparticle is toxic in nature. It basically evaluates the nature and properties of NP to find out the severity of toxic nature of these particles individually, as well as in combination with other particles or compounds. The techniques of nanotoxicity prediction include the experimental assessments of the nanocompounds which are performed in the laboratories in a specific environment. Another method is testing of the particle or compound on living organisms. Both of these procedures are costly and time-consuming. An alternate approach for the same is to perform computational modelling on the available previous experimental data. Along with experimental data, various physicochemical properties or theoretical molecular descriptor values can also be utilized in the toxicity prediction. This process is known as quantitative

© The Author(s) 2019
Ch. Satyanarayana et al., *Computational Intelligence and Big Data Analytics*,
SpringerBriefs in Forensic and Medical Bioinformatics,
https://doi.org/10.1007/978-981-13-0544-3_6

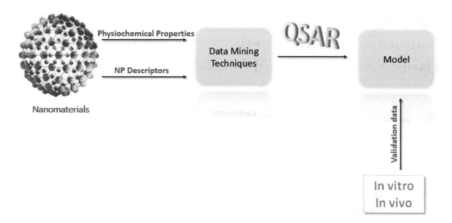

Fig. 6.1 QSAR model

structured–activity relationship (QSAR). QSAR is a mathematical model used to find out the reactivity level of different compounds based on their properties, descriptors, and previous experimental data [1, 2]. As shown in Fig. 6.1, QSAR model takes physicochemical properties and NP descriptors as input and process them using various data mining techniques. Further, the predicted data is validated with actual experimental values. In the next section, we have discussed the various ways to predict the toxicity of nanomaterials using data mining techniques.

6.2   Identification of Properties

6.2.1   Physicochemical Properties

The first and foremost requirement of QSAR is to identify the appropriate physicochemical properties on which the toxicity of a compound depends. An extensive list of significant physicochemical characteristics has been provided by OECD [3] for toxicology analysis. Here, we are discussing the basic properties on which the toxicity evaluation was performed earlier.

(1) **Size**: Size of a particle plays an important role in generating the toxic nature of a molecular compound. The smaller the size of a nanoparticle, the more easily it can invade through the cells and can be harmful. Various toxicity evaluations were done in the past based on size value of nanoparticles [4–7]. To make these studies more precise, size distribution was considered as quantifier in QSAR. The previous studies found that smaller size nanomaterials are more toxic as compared to large size.

(2) **Shape**: Shape is another important factor which plays an important role in generating toxicity in a compound. The quantifiers of shapes includes spherical, rectangle, long, short, rod-like [5, 8].

(3) **Surface characteristics**: Surface of the nanoparticles contains various features such as coating and charges (positive, negative, and neutral) which affects the toxicity of a molecule [9–11]. In previous studies, it was found that negatively charged compounds are more prone to toxic nature as compared to less negative. Similarly, positively charged compounds are more toxic as compared to neutral one [12–14].

(4) **Crystal structure**: Crystal structure plays an important role in determining the toxicity, as the compound with same composition can change its toxic nature with a change in its crystal structure [15, 16].

6.2.1.1 Outcomes and Discussion

The previous studies imply that the physicochemical properties are the possible and significant factors which are partly responsible for the toxic nature of a compound. As mentioned above, with the decrease in particle size, the toxicity level gets increases. In this case, the data of size distribution matters a lot; as on application of varied techniques such as particle aggregation, the measured size gets increase. The adoption of correct measurement techniques helps in accurate prediction.

The studies based on shape figure out that long-shaped particles are more toxic as compared to short one. The molecules with same composition can have different shape, which can vary the toxicity level also. Here, again the shape measurement techniques matter a lot. This area needs the state-of-the-art techniques to explore these properties for toxicity prediction.

Another crucial factor which impacts the toxicity is surface functionality. The surfaces of nanomaterials sometimes get coated deliberately and undeliberately to have required functionality. However, sometimes these coatings can result in variations in aggregation state and size of particles, which further affects the toxicity. As per previous studies [12–14], the uncoated surface is less toxic as compared to coated.

Other than these, there are various physicochemical properties, which contribute in generating toxic nature of compounds. Exploring more physicochemical properties for toxicity prediction is an emerging research area and needs a lot more work to be done.

6.2.2 Theoretical Chemical Descriptor

Apart from basic properties, there are various theoretical descriptors, which helps in finding out the reactivity level of compounds. Initially, to find out the values of these descriptors, the electronic structure calculation of a compound is done on the basis of density functional theory. All these calculations are done using some softwares or

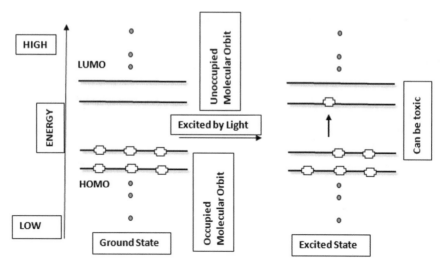

Fig. 6.2 HOMO–LUMO gap and reactivity in molecular state

tools. While working on molecular orbitals, the two values HOMO and LUMO are identified.

HOMO: Highest occupied molecular orbital,
LUMO: Lowest energy unoccupied molecular orbital.

The difference between these is known as HOMO–LUMO gap as shown in Fig. 6.2, which is the lowest energy excitation that is possible in a molecule [17]. Based on this, we can predict the toxicity of a molecule. The molecules with more HOMO–LUMO gap are not much reactive. While on the other hand, the tightly coupled molecules with less HOMO–LUMO gap have high probability of become toxic. This gap allows the molecular orbit to move easily from low energy level to high energy level. This change in molecular data can make it harmful.

Initially using some software tools, the molecular structure is explored and theoretical values of HOMO–LUMO are found out. There are various molecular descriptors which depend on these values; hence using the HOMO–LUMO value, we can also find the other descriptors values. Few of these descriptors are mentioned below [18]:

Molecular hardness: It is described as the variation in chemical potential μ upon variation in the number of electrons N, which can be approximated with the values of HOMO and LUMO.

$$\eta = -\left(\frac{\partial \mu}{\partial N}\right) \approx -(E_{\text{HOMO}} - E_{\text{LUMO}}) \qquad (6.1)$$

Molecular softness: Molecular softness also derives the changes produced.

$$S = \frac{1}{2\eta} \qquad (6.2)$$

Selectivity–reactivity descriptor: This descriptor is used to measure the reactivity toward the nucleophilic and electrophilic attacks. Here, ρ_{HOMO} and ρ_{LUMO} are defined as electron densities of HOMO and LUMO orbitals.

$$\Delta f(r) \approx \rho_{LUMO}(r) - \rho_{HOMO}(r) \tag{6.3}$$

Electrophilicity index: This is one of the popular descriptors used in the toxicity prediction at the molecular level. Here, χ is defined as molecular electrophilicity and η as molecular hardness. This value can also be assessed using HOMO and LUMO values.

$$\omega = \frac{\chi^2}{2\eta} = \frac{E_{HOMO}^2 + 2E_{HOMO}E_{LUMO} + E_{LUMO}^2}{4(E_{LUMO} - E_{HOMO})} \tag{6.4}$$

In [19], author has used the electrophilicity index as a descriptor and analyzed the relationship between toxicity nature and its value. Other than these, there are various other descriptors which can be evaluated in QSAR using these HOMO and LUMO values and can facilitate nanotoxicity forecasting.

6.3 Computational Techniques

The previous section discusses the various ways to evaluate toxic nature of a compound. To explore the available data, various computational or data mining techniques are required to smooth the evaluation process. In [2, 20–22], the authors have reviewed the QSAR approaches applied in the past. The next section discusses the most widely used techniques in the prediction of nanotoxicity.

Regression: Regression analysis is widely used in prediction and forecasting. It explains the variation occurring in value of dependent variables with respect to independent variable. The most widely used linear regression algorithms include multiple linear regression (MLR), partial least square regression (PLSR), and principle component regression (PCR) [23–25]. The PLS regression is used when there is intercorrelation between certain descriptors. These techniques prove beneficial when multiple responses needed to be modeled concurrently. The foremost advantage of linear regression methods is their transparency to provide relevant significance information of physicochemical descriptors. Sometimes, these models prove to be poorly performed because of the presence of input variables which are besides the output. Hence, to remove these various dimension reduction techniques such as principle component analysis (PCA) is used. Various past studies had used PCA to improve the outcome of their QSAR model.

Decision tree: A decision tree is an efficient approach used in classification and regression. In QSAR, it can be used to predict toxicity on the basis of numerical and categorical data of biological activities [26, 27]. The significant capabilities of decision tree which contribute the most in toxicity prediction include automatic selection of input variables, removal of insignificant descriptors, and to find worthwhile out of small, noisy, and large datasets.

Support vector machine: SVM proves to be an alternate approach for linear modeling which can manage both classification and regression problems [28, 29]. It can handle numerous issues such as collinear descriptors, nonlinear relations, small and large datasets as well as over-fitted models. The potential of SVM to provide high accuracy and generalization makes it an approach of interest.

Artificial neural network: ANN is another approach used for nonlinear data relationship and huge datasets. ANN approach has been applied in past but has certain disadvantages such as choosing the ideal complexity, trouble of overfitting, high generalization sensitivity to variation in parameters and network topology, and, last but not least, difficulty in predicting the outcome [30, 31].

6.4 Prediction on the Basis of Live Cells

As nanomaterials are being used in drug delivery systems, there is a great need to find out the intensity of harmfulness, which can arise due to the association of different materials. Whenever the drug formation takes place, the sample of it is checked microscopically. The microscopic images generated are further analyzed to check its toxic nature, which is decided on the basis of cell count. After the drug loading process, large number of live cells gets dead, and then it becomes toxic. If there is no change in the numbers of live cell, it is safe to use. Image processing techniques can prove effective in determining the status of cells. Using these microscopic images as input and image processing techniques, the live and dead cells can be counted precisely.

6.5 Experimental Analysis

In this paper, regression analysis was applied to find out the variation in toxicity nature of a compound with respect to electrophilicity index. As mentioned in Sect. 6.2.2, electrophilicity index value as a theoretical descriptor can be derived from HOMO–LUMO values. Hence, initially we have found out the HOMO–LUMO values for testosterone derivatives and further derived the electrophilicity index value out of it using Eq. 6.4.

As discussed earlier in Sect. 6.4, the regression analysis is used for finding out the best-fitted linear relationship between two variables. Therefore, we have applied the parametric regression to find the linear relationship between electrophilicity index and other biological parameter values. Table 6.1 shows the result of analysis done. The results predicted were good and quiet identical to experimental values.

Table 6.1 Relationship of electrophilicity index with various biological activities of testosterone derivatives

S. No.	Biological activity name	R & SD (Experimental)	R & SD (Calculated)
1.	Relative Binding Affinity (RBA)	$R = 0.899$, SD $= 0.236$	$R = 0.975$, SD $= 0.096$
2.	Therapeutic index (TI)	$R = 0.781$, SD $= 0.067$	$R = 0.773$, SD $= 0.068$
3.	TeBG affinity (TeBG)	$R = 0.612$, SD $= 0.644$	$R = 0.615$, SD $= 0.642$
4.	Relative competition indices (RCIs)	$R = 0.845$, SD $= 0.094$	$R = 0.843$, SD $= 0.094$
5.	Prostate receptor protein (PRP)	$R = 0.716$, SD $= 0.835$	$R = 0.723$, SD $= 0.25$
6.	Myotro Androgenic Potency Temp (MAPT)	$R = 0.883$, SD $= 0.121$	$R = 0.881$, SD $= 0.123$

R residual error, *SD* standard deviation

6.6 Affirmation of the Model

A QSAR model needs to fulfill some elementary quality assessments suggested by OECD, to validate its findings. Serving all these regulations is really very challenging. One of the reasons behind this is the variations in the protocols followed, which as a result gets reflected in the outcome of the experimental data. Another major problem faced in validating the model is the certainty of the algorithm used for nanostructure data. To run over this problem, adequate descriptors are used along with previously available data.

6.7 Conclusion

With the increase in number of engineered nanomaterials, the acceptance of hazardous nature of these materials to human health has also been increased. For this, it is required to evaluate the toxic nature of these materials as well as their variants. In order to perform toxicity checking and parallel avoiding animal testing, QSAR proves to be an emerging alternative approach. QSAR models need to be explored more along with various data mining techniques which can be applied in the prediction process. The paper has illustrated the QSAR process, the data, descriptors as well as the various mining methods used in past and has also shown some experimental results based on regression analysis. This paper provides the research direction to work in this area.

References

1. Maojo V, Fritts M, de la Iglesia D, Cachau RE, Garcia-Remesal M, Mitchell JA, Kulikowski C (2012) Nanoinformatics: a new area of research in nanomedicine. Int J Nanomed 7:3867–3890. https://doi.org/10.2147/IJN.S24582 PMID:22866003
2. Oksel C, Ma CY, Liu JJ, Wilkins T, Wang XZ (2017) Literature Review of (Q) SAR modelling of nanomaterial toxicity. In: Modelling the toxicity of nanoparticles. Springer International Publishing, Basel, pp 103–142
3. OECD (2010) guidance manual for the testing of manufactured nanomaterials: OECD's Sponsorship Programme. Organization for Economic Co-operation and Development, Paris
4. Pettitt ME, Lead JR (2013) Minimum physicochemical characterisation requirements for nanomaterial regulation. Environ Int 52:41–50
5. Powers KW, Palazuelos M, Moudgil BM, Roberts SM (2007) Characterization of the size, shape, and state of dispersion of nanoparticles for toxicological studies. Nanotoxicology 1(1):42–51
6. Park MV, Neigh AM, Vermeulen JP, de la Fonteyne LJ, Verharen HW, Briedé JJ et al (2011) The effect of particle size on the cytotoxicity, inflammation, developmental toxicity and genotoxicity of silver nanoparticles. Biomaterials 32(36):9810–9817
7. Karlsson HL, Gustafsson J, Cronholm P, Möller L (2009) Size-dependent toxicity of metal oxide particles a comparison between nano-and micrometer size. Toxicol Lett 188:112–118
8. Gratton SE, Ropp PA, Pohlhaus PD, Luft JC, Madden VJ, Napier ME, DeSimone JM (2008) The effect of particle design on cellular internalization pathways. Proc Natl Acad Sci 105:11613–11618
9. Zhao CM, Wang WX (2012) Importance of surface coatings and soluble silver in silver nanoparticles toxicity to Daphnia magna. Nanotoxicology 6(4):361–370
10. Park YH, Bae HC, Jang Y, Jeong SH, Lee HN, Ryu WI et al (2013) Effect of the size and surface charge of silica nanoparticles on cutaneous toxicity. Mol Cell Toxicol 9(1):67–74
11. Bhattacharjee S, de Haan LH, Evers NM, Jiang X, Marcelis AT, Zuilhof H (2010) Role of surface charge and oxidative stress in cytotoxicity of organic monolayer-coated silicon nanoparticles towards macrophage NR8383 cells. Part Fibre Toxicol 7(1):25
12. Caballero-Díaz E, Pfeiffer C, Kastl L, Rivera-Gil P, Simonet B, Valcárcel M et al (2013) The toxicity of silver nanoparticles depends on their uptake by cells and thus on their surface chemistry. Part Part Syst Charact 30(12):1079–1085
13. Nguyen KC, Seligy VL, Massarsky A, Moon TW, Rippstein P, Tan J, Tayabali AF (2013) Comparison of toxicity of uncoated and coated silver nanoparticles. In: *Journal of Physics: Conference Series,* vol 429, No 1. IOP Publishing, Bristol, p 012025
14. Yang X, Gondikas AP, Marinakos SM, Auffan M, Liu J, Hsu-Kim H, Meyer JN (2011) Mechanism of silver nanoparticle toxicity is dependent on dissolved silver and surface coating in Caenorhabditis elegans. Environ Sci Technol 46(2):1119–1127
15. Jiang J, Oberdörster G, Elder A, Gelein R, Mercer P, Biswas P (2008) Does nanoparticle activity depend upon size and crystal phase? Nanotoxicology 2(1):33–42
16. Napierska D, Thomassen LC, Lison D, Martens JA, Hoet PH (2010) The nanosilica hazard: another variable entity. Particle Fibre Toxicol 7(1):39
17. Zhang H, Ji Z, Xia T, Meng H, Low-Kam C, Liu R, Nel A et al (2012) Use of metal oxide nanoparticle band gap to develop a predictive paradigm for oxidative stress and acute pulmonary inflammation. ACS Nano 6(5):4349–4368. https://doi.org/10.1021/nn3010 087 PMID:22502734
18. Novel Descriptor for Reactivity http://cordis.europa.eu/documents/documentlibrary/1168104 41EN6.pdf
19. Parthasarathi R, Subramanian V, Roy DR, Chattaraj PK (2004) Electrophilicity index as a possible descriptor of biological activity. Bioorg Med Chem 12(21):5533–5543
20. Sizochenko N, Leszczynski J (2016) Review of current and emerging approaches for quantitative nanostructure-activity relationship modeling: the case of inorganic nanoparticles. J Nanotoxicol Nanomed (JNN) 1(1):1–16

21. Yosefu NO (2015) Computational modelling for prediction of nanomaterial toxicity. Doctoral dissertation, Makerere University
22. Oksel C, Ma CY, Liu JJ, Wilkins T, Wang XZ (2015) (Q) SAR modelling of nanomaterial toxicity: a critical review. Particuology 21:1–19
23. Gu C, Goodarzi M, Yang X, Bian Y, Sun C, Jiang X (2012) Predictive insight into the relationship between AhR binding property and toxicity of polybrominated diphenyl ethers by PLS-derived QSAR. Toxicol Lett 208:269–274
24. Shahlaei M (2013) Descriptor selection methods in quantitative structure? activity relationship studies: a review study. Chem Rev 113:8093–8103
25. Yee LC, Wei YC (2012) Current modeling methods used in QSAR/QSPR. Assessment 10:11
26. Bengio Y, Delalleau O, Simard C (2010) Decision trees do not generalize to new variations. Comput Intell 26:449–467
27. Wang XZ, Ma CY (2009) Morphological population balance model in principal component space. AIChE J 55:2370–2381
28. Darnag R, Minaoui B, Fakir M (2012) QSAR models for prediction study of HIV protease inhibitors using support vector machines, neural networks and multiple linear regression. Arab J Chem
29. Mei H, Zhou Y, Liang G, Li Z (2005) Support vector machine applied in QSAR modelling. Chin Sci Bull 50:2291–2296
30. Sussillo D, Barak O (2013) Opening the black box: low-dimensional dynamics in high- dimensional recurrent neural networks. Neural Comput 25:626–649
31. Ventura C, Latino DA, Martins F (2013) Comparison of multiple linear regressions and neural networks based QSAR models for the design of new antitubercular compounds. Eur J Med Chem 70:831–845

Chapter 7
Stock Market Prediction Based on Machine Learning Approaches

V. Lalithendra Nadh and G. Syam Prasad

Abstract Forecasting stock market based on the information available with high precision is not so consistent because of its unsteady nature. There are numerous approaches in the anticipation of stock markets. Machine learning systems are a standout among other methodologies in expectation. Numerous researchers have done wide research over the years using different machine learning algorithms. In this paper, the written work examines on different computational tools such as genetic algorithms (GAs), support vector machine (SVM), artificial neural networks (ANNs) are used for stock market forecasting.

Keywords Genetic algorithms (GAs) · Support vector machine (SVM)
Artificial neural networks (ANNs)

7.1 Introduction

Forecasting of a stock market is an uncertain undertaking in the business environment. The fundamental standard of a stock market is price rising of a share based on the increase in investment in business and the other way around. Forecasting the stock price may prompt benefit or misfortune that the trader needs to tolerate. Any anticipation will help in decision making either to buy or sell that particular share. Based on several reliable anticipation strategies and news items, numerous traders are trading by compelling some calculated risks. The informative news from various media channels has a high impact on share value movement. Fundamental analysis is enough for investors. Technical analysis and short-term Predictions are required for trading. Many inputs are to be considered to anticipate the future value in the stock market. But it is absolutely confusing the traders while taking decisions. The successful formula for the anticipation of a stock market is getting the finest outcomes with less number of data sources. Forecasting is very difficult due to its volatile nature. The movement average of various stocks was incorporated in the stock market index. Market movement is reflected by the index rather than an individual stock movement. To such an extent, numerous researchers focused on forecasting

© The Author(s) 2019
Ch. Satyanarayana et al., *Computational Intelligence and Big Data Analytics*,
SpringerBriefs in Forensic and Medical Bioinformatics,
https://doi.org/10.1007/978-981-13-0544-3_7

individual stock prices. There are a few methodologies that are utilizing machine learning algorithms. The most well-known approaches are genetic algorithms (GAs), support vector machine (SVM), artificial neural networks (ANNs).

7.2   Literature Review

Kyoung-jae Kim [1] proposed a genetic algorithm (GA) for the expectation of securities exchange in an unexpected way. The main advantage of GA is to propel the learning calculation and many-sided quality diminishment in feature space. Kyoung-jae Kim [1]. Authors proposed GA in a different way to calculate the affiliation weights for feature discretization (FD) and for ANN during the forecast of stock value. In general, for getting the affiliation weights in ANN, gradient descent algorithm is utilized, and backpropagation algorithm is utilized as a local search algorithm. According to [1], author point of view, gradient descent algorithm performs very poorly when contrasted with GA. The learning procedure will be simplified when the information is appropriately discretized, and learned consequences of enhanced generalizability will diminish the redundant and noisy information adequately. Prior to the search process, the affiliation weights and inception for feature discretization are initialized to arbitrary values. The genetic algorithm feature discretization (GAFD) in Fig. 7.1 has three stages mainly. Three sets of parameters have been taken in the primary stage. To start with a set of affiliation weights in the middle of the information layer and hidden layer of the system, the second set affiliation weights in the middle of a concealed layer and yield layer are edges for FD in the third set. The searching parameters are encoded on chromosomes to influence the best utilization of the fitness function. The determined affiliation weights with feedforward computation from the primary stage and the linear function are consolidated in the second stage. The thresholds for FD and derived affiliation weights are valuable to the proposed information in the final stage.

Kim [2] proposed expectation of consistent value change in stock record utilizing support vector machine (SVM). SVM is a sort of learning algorithm in Fig. 7.2 that locates a unique kind of highest margin hyperplane and linear model. Support vectors are only the preparation illustrations that are near to highest margin hyperplane. The binary decision classes are isolated by a hyperplane for the linearly divisible element. Deciding the parameters and finding the support vectors are nothing but unraveling a quadratic program which is linearly compelled. SVM alters the contributions of high-dimensional component space to normalize nonlinear class by building a direct model. Among different parts, the best model must be picked which limits the estimate. Polynomial kernel and Gaussian outspread basis function are the normal cases of a kernel function. If the two parameters are chosen inadequately, it will cause under-fitting or over-fitting issues. The information factors that are utilized as a part of this approach are Korea Composite Stock Price Index (KOSPI) on a regular schedule and technical indicators. The traditional change in stock price index is delegated as "1" or "0". "0" denotes that today's index is prominent than following

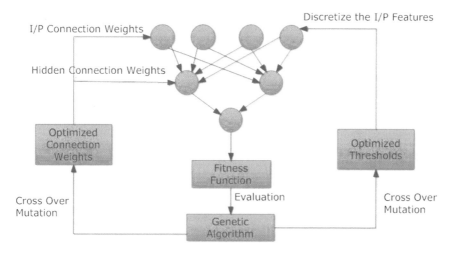

Fig. 7.1 Framework of genetic algorithm feature discretization

Fig. 7.2 Support vector machine

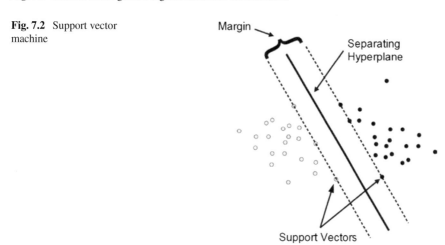

day, whereas "1" denotes that today's index is lesser than the following day. Every feature component is standardized independently to the specific range through linear scaling which ensures that greater value inputs do not amaze inputs of smaller value, to reduce errors during forecasting. To recover similar cases, the closest neighbor strategy is utilized for case-based reasoning (CBR). The closest neighbor strategy is normal recovery technique which can be simply applied to financial information. Euclidean distance is the assessment function of closest neighbor strategy. At last, the outcomes were satisfied when contrasted with backpropagation network.

de Oliveira and Nobre [3] proposed a technique for stock market forecasting using artificial network approach. This approach comprises three stages in Fig. 7.3 to anticipate the tendency and behavior of a stock. The three stages are (1) getting samples,

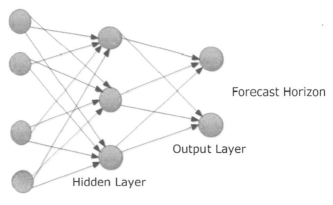

Forecast Horizon

Output Layer

Hidden Layer

Input Layer

Fig. 7.3 Artificial neural network structure

(2) input preprocessing, and (3) forecast. The foremost objective is to anticipate the closing price of a particular stock. Samples were taken from a stock named PETR4 which trades on BM&FBOVESPA. The preprocessing was done on day-by-day quotes of that specific stock together with some technical indicators. The technical indicators are nothing but mathematical computations which will be connected to a series of information. For each date, if there is any misplaced information, that specific date ought to be erased from the sample set. The linear intersection strategy was picked during the process of normalization for a series of sample sets having different esteem scales. One of the machine learning tasks called supervised learning was chosen for investigating the training data. The backpropagation algorithm was chosen for training the network to calculate the error contribution. Windowing process is utilized to pass the sample sets in addition with forecast horizon and size of the window picked. At last, the training process was finished and the performance was measured utilizing validation and test.

7.3   Conclusion

For any trader or investor, the stock market forecast will help in making appropriate decisions and furthermore gives some clarity during buying and selling the stocks. This paper is reviewed on a stock market forecast by utilizing the genetic algorithms (GAs), support vector machine (SVM), artificial neural networks (ANNs). The behavior and trends of the stock are explained, the closure of the next day is the reference function, the five-day annotations Successfully obtaining appropriate results. ANN mainly focused on improving the learning algorithm and association weights. Combination of genetic algorithm with ANN has been proposed to overcome the limitation of ANN. GA looks for close optimum thresholds of feature discretization and close to best possible solutions of association weights in learning algorithm. The study of the genetic algorithm also has a few constraints while processing the elements in hidden layer and optimization of a learning process. To overcome the ordinary procedures, SVM anticipated structural risk minimization standard for the enhanced model. SVM got the better accurate outcomes contrasted with the back-propagation in neural networks. Each algorithm has some advantages as well as few limitations. Hence, the combination of some technical indicators and the machine learning algorithms will give best results.

References

1. Kyoung-jae Kim IH (2000) Genetic algorithms approach to feature discretization in artificial neural networks for the prediction of stock price index, pp 125–132, Elsevier
2. Kim K-J (2003) Financial time series forecasting using support vector machines, pp 307–319, Elsevier
3. de Oliveira FA, Nobre CN (2011) The use of artificial neural networks in the analysis and prediction of stock prices. IEEE

Chapter 8
Performance Analysis of Denoising of ECG Signals in Time and Frequency Domain

CH. Hima Bindu

Abstract This paper addressing various denoising techniques of electrocardiogram signal (ECG) with basic filters both in time and frequency domain. The electrocardiogram (ECG) signal is vital for accurate diagnosing of heart disease. ECG modeling and noise reduction are rather essential for clinical applications also. The major issue is denoising of ECG signal which has been corrupted by equipment or by transmission process. Several methods have been applied for modeling and denoising of ECG signals with low-, high-, band-, notch-pass filters, etc. These methods demonstrated good performance; they can be sensitive to varying parameters. Denoising in frequency domain, the signal is transformed in the discrete wavelet transform domain, where thresholding (soft and hard thresholding)-based noise reduction algorithm is employed. Finally, the calculation of different performance measures enables us to choose an efficient technique. The accuracy and consistency of the proposed method are shown in experimental results.

Keywords Denoising · Time domain filters · Frequency domain filter: DWT

8.1 Introduction

Electrocardiogram (ECG) is the tool of electrodes used to record electrical potential of the heart over interval of time. These electrodes are responsible to detect electrical changes happens from the heart muscles. It is usually used for cardiology test to diagnosis the functioning of heart. The ECG output which is traced as PQRST waves is shown in Fig. 8.1. The careful observation of these waves can avoid the chance of occurrence of chronic diseases ultimately. This tool can be used to measure the heartbeat rate, size, and position of the heart chambers.

Gupta et al. [10] developed an embedded system for acquisition and analysis of ECG signal. Bui et al. [4] discussed e-processing of ECG data. Zewang et al. [5] denoising of ECG with adaptive Fourier decomposition. His work is validated with standard database ECG signals. Nguyen et al. [7] proposed a methodology for removal noise in ECG signal with adaptive denoising, EEMD, and genetic algorithm-

© The Author(s) 2019
Ch. Satyanarayana et al., *Computational Intelligence and Big Data Analytics*,
SpringerBriefs in Forensic and Medical Bioinformatics,
https://doi.org/10.1007/978-981-13-0544-3_8

Fig. 8.1 Representation of
ECG signal

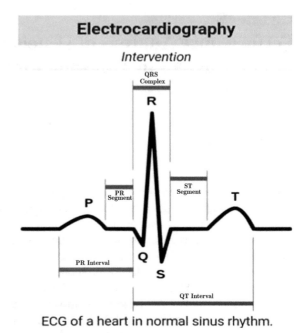

ECG of a heart in normal sinus rhythm.

based thresholding techniques. Xiong et al. [8] proposed contractive denoising process to enhance the performance of auto-encoders for ECG denoising. Das et al. [9] discussed filtering of ECG signal with S-transform.

The wavelet transform can effectively separate the signal and noise. In biomedical signal processing the wavelet transform is mainly applicable for signal compression, feature extraction and signal denoising [2, 4]. The performance of signal processing algorithms can be evaluated with Physionet [6] database.

8.2   Denoising

Noise is defined as any unpleasant or unexpected signal coming from some sources, which interferes the expected signal [1]. This noisy signal affects the quality of the signal. Noise has to be neglected when performing further analysis. Denoising of signal can be defined as a process for separating a signal mixed with noise.

The perfect processing and signal analysis can be done once the noise is removed from the signal. This can make possible with denoising algorithms. The denoising process should retain the original signal without disturbing its quality [3]. The traditional way of denoising is with low- or band-pass filters. Hence, it is giving the scope for ECG denoising for perfect clinical diagnosis.

8.3 Denoising Filters

The following are the different time domain filters used in the proposed algorithm to reconstruct the original signal [11].

- Butterworth filters
- Chebyshev type I filter
- Chebyshev type II filter
- IIR notch filter
- Elliptical filter

(i) Butterworth filter:

The Butterworth ideal low-pass filter response offers the Taylor series approximation at analog frequencies $w = 0$ and $w = \infty$. Response is monotonic overall, decreasing smoothly from to $w = 0$ and $w = \infty$. The transfer function is (Fig. 8.2):

$$H(jw) = \frac{1}{\sqrt{2}} \text{ at } w = 1 \tag{8.1}$$

(ii) Chebyshev type I filter:

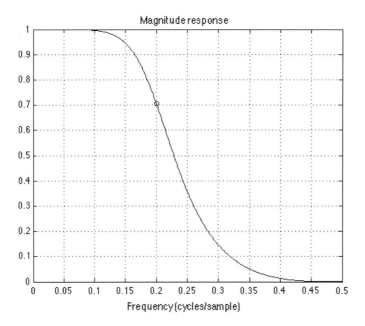

Fig. 8.2 Frequency response of Butterworth filter

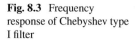

Fig. 8.3 Frequency
response of Chebyshev type
I filter

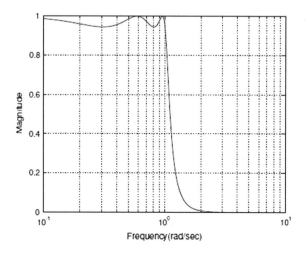

The Chebyshev type I filter suppresses the difference between the ideal and actual frequency responses in pass band. It incorporates a ripple of 0Rp dB in this band. The response is flat in stop band.

$$H(jw) = \frac{1}{\sqrt{1 + \epsilon^2\, T_n^2\left(\frac{w}{w_0}\right)}} \tag{8.2}$$

where ϵ is the ripple factor, w_0 is the cutoff frequency, and T_n is a Chebyshev polynomial of the nth order (Fig. 8.3).

(iii)   Chebyshev type II filter:

The Chebyshev type II filter introduces ripples in the stop band to suppress difference in ideal and actual frequency responses. The frequency response of Chebyshev type II filter is flat at pass band response and ripples at stop band. The stop band magnitude is not set to zero as type-I filter (Fig. 8.4).

$$H(jw) = \frac{1}{\sqrt{1 + \epsilon^2\, T_n^2\left(\frac{w_0}{w}\right)}} \tag{8.3}$$

(iv)   Elliptical filters:

Elliptic filters are equiripple in both the pass band. It is a combination of Chebyshev and Taylor series in both pass and stop bands. This filter is named as elliptic filter due to the usage of elliptic function for calculation of pole and zero locations. It is also called rational Chebyshev or Cauer filter. Elliptic filters also minimize the transition width.

Fig. 8.4 Frequency
response of Chebyshev type
II filter

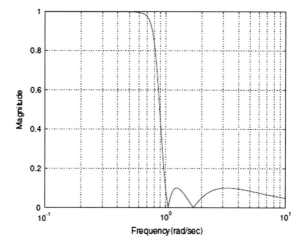

Fig. 8.5 Frequency
response of elliptical filter

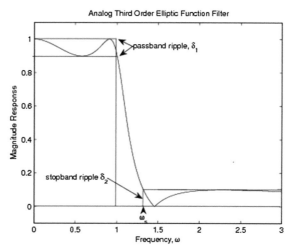

$$H(jw) = \frac{1}{\sqrt{1 + \epsilon^2 \, G(w)^2}} \qquad (8.4)$$

where $G(w)$ is defined by Jacobian elliptic functions and ϵ pass band ripple (Fig. 8.5).

(v) Notch filter:

Notch filter is an elimination of certain frequency of the signal or allowing of specific frequency of the signal. The ideal notch filter frequency response of the above cases is mentioned below (Fig. 8.6):

Fig. 8.6 Frequency
response of notch filter

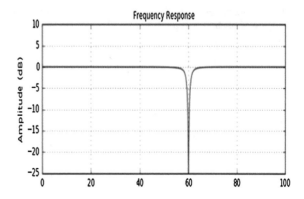

$$H(jw) = \begin{cases} 0 & \omega_0 \\ 1 & \text{otherwise} \end{cases}$$

$$H(jw) = \begin{cases} 1 & \omega_0 \\ 0 & \text{otherwise} \end{cases} \qquad (8.5)$$

8.4 Proposed Algorithm in Time Domain

Denoising in time domain results the usage of different filtering techniques. The
algorithm for signal denoising in time domain is mentioned as follows and shown in
Fig. 8.7:

INPUT: Noisy ECG signal
OUTPUT: De-noised ECG signal

1. Reading an ECG signal $X(n)$ from the database with sampling frequency **Fs** and
   sampling period **Ts**.
2. Applying additive white Gaussian noise $N(n)$ on ECG signal with various densities.

$$Y(n) = X(n) + N(n) \qquad (8.6)$$

3. Performing denoising using various filters like Butterworth, Chebyshev type I
   and II, IIR notch, and elliptical low-pass filters $H(jw)$.

$$D(n) = Y(n) * H(jw) \qquad (8.7)$$

Fig. 8.7 Proposed flow
chart of time domain filtering

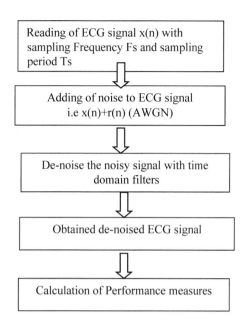

4. The response of the filter is the denoised ECG signal D(n).
5. Comparing the performance of the each filter by measuring: PSNR, SNR, and MSE.

8.5 Denoising in Frequency Domain

(i) Wavelet transform

The wavelet transform converts time domain information into frequency domain. The signal can be analyzed more conveniently in frequency domain rather than in time domain. Conventional filtering is not suitable for the signal whose frequency band is similar to noise frequency band. Therefore, wavelet-based methods are useful, and here, the signal is decomposed into frequency bands (scales). Different types of wavelets can be used while decomposing signal into DWT coefficients [12].

Figure 8.7 illustrates basic wavelet functions (a) Haar, (b) Daubechies, (c) coiflet 1, (d) symlet 2, (e) Meyer, (f) Morlet, (g) Mexicanhat. Haar wavelet is the oldest and simplest wavelet rather than rest of the functions.

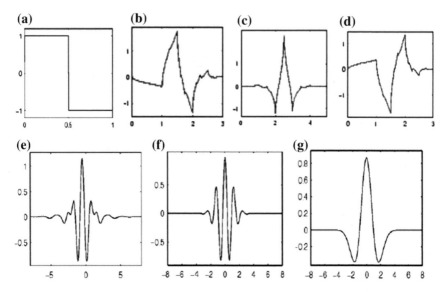

Fig. 8.8 Different types of wavelets

8.6   Proposed Algorithm in Frequency Domain

The frequency domain denoising algorithm follows the steps described below and shown in Fig. 8.8:

1. Reading of ECG signal $x(n)$ with sampling frequency F_s and sampling period T_s.
2. Adding additive white Gaussian noise to the ECG signal, i.e., $x(n) + r(n)$ is resulting noisy signal.
3. Apply DWT on above signal. The resulting coefficients are approximation and detail coefficients.
4. Perform thresholding to the DWT coefficients by selecting proper threshold value. Two types of thresholding can applied

   • soft thresholding, and
   • hard thresholding.

5. Applying inverse discrete wavelet transform (IDWT) to the coefficients to obtain denoised signal, i.e., $x(n)$.
6. Calculating performance measures of denoised signal to both soft thresholding and hard thresholding results.

Table 8.1 Performance measures of all filters both in time and frequency domains

Filtering technique	MSE	SNR (db)	PSNR (db)	Cross-correlation
IIR notch filter	0.0049	6.57	19.28	0.88
Chebyshev type I filter	0.0002	19.26	31.96	0.993
Elliptical filter	0.0001	20.545	33.24	0.996
Butterworth filter	0.00001	31.58	44.29	0.999
Chebyshev type II filter	0.00009	34.17	46.87	0.999
DWT by applying thresholding	0.000002	41.05	53.76	0.99

8.7 Results and Discussion

This section deals with the experimental results of the proposed work. The signals are collected from [6] Web link. Performance of denoising techniques among various filters is compared and tabulated in Table 8.1. The used performance measures are signal-to-noise ratio (SNR), mean square error (MSE), peak signal-to-noise ratio (PSNR), and cross-correlation [13]. These are discussed below:

(i) Signal-to-noise ratio:

Signal-to-noise ratio is the ratio of signals to the noise powers. It is used to verify the performance of the signal quality. It is expressed in decibels.

$$SNR = 20 * \log \frac{P_s}{P_w} \tag{8.8}$$

$$SNR = \frac{\sum_{t=0}^{L-1} s^2(t)}{\sum_{t=0}^{L-1} n^2(t)} \tag{8.9}$$

The signal and noise powers must be measured at the same equivalent points and at same bandwidth.

(ii) Mean square error:

MSE is defined as sum of squares of the signal.

$$MSE = \frac{1}{N \sum_{n=1}^{N} \left(X[n] - \hat{X}[n] \right) 2} \tag{8.10}$$

MSE should be small for better denoising technique.

(iii) Peak signal-to-noise ratio:

Fig. 8.9 Proposed denoising algorithm in time domain

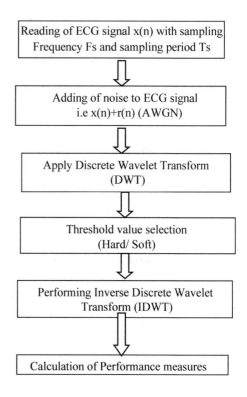

PSNR means peak signal-to-noise ratio, and it is the ratio of maximum signal strength to the mean square error. PSNR is usually expressed in terms of the logarithmic decibel scale.

$$PSNR = 10 * \log_{10}\left(\frac{MAX_I^2}{MSE}\right)$$

$$= 20 * \log_{10}\left(\frac{MAX_I}{\sqrt{MSE}}\right) \quad (8.11)$$

(iv) Cross-correlation:

The cross-correlation is a measure of similarity of two signals as a function of the displacement of one relative to the other (Figs. 8.9, 8.10, 8.11, 8.12, 8.13, 8.14, 8.15, and 8.16).

$$r_{xy}(h) = \begin{cases} \sum_{n=0}^{N-h-1} X(n+h)Y^*(n) & 0 \le h \le N-1 \\ r_{yx}^*(h) & -(N-1) \le h \le 0 \end{cases} \quad (8.12)$$

The following table displays the details of the performance measures of various algorithms among time and frequency domain techniques.

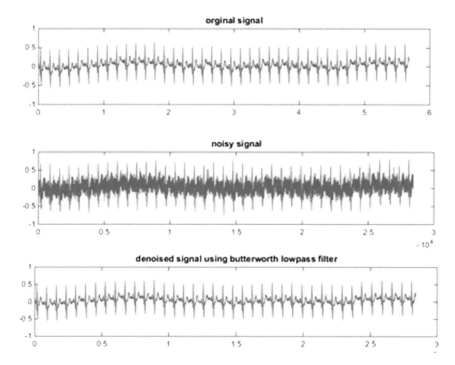

Fig. 8.10 Denoised signals with Butterworth filter

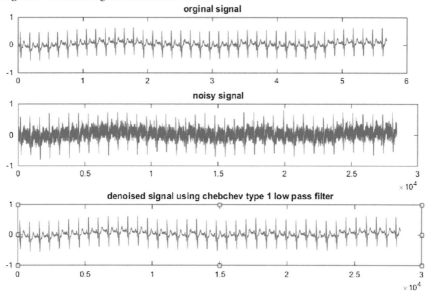

Fig. 8.11 Denoised signals with Chebyshev I filter

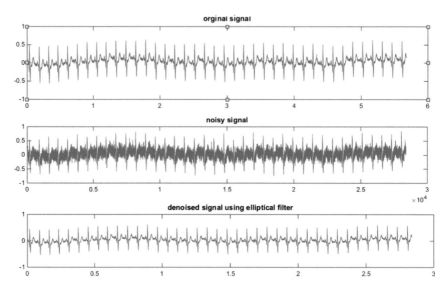

Fig. 8.12 Denoised signals with Chebyshev II filter

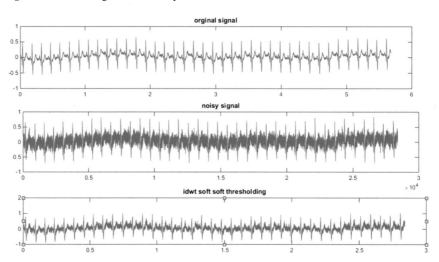

Fig. 8.13 Denoised signals with elliptical filter

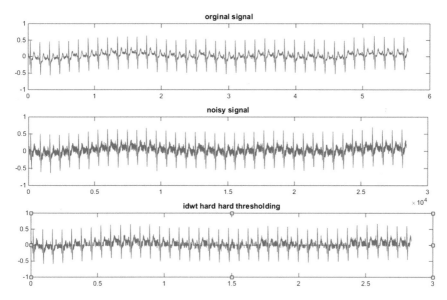

Fig. 8.14 Denoised signals with IIR notch filter

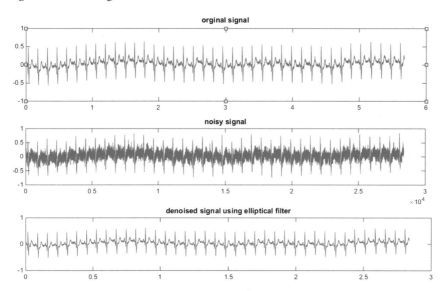

Fig. 8.15 Denoised signals with soft threshold signal to DWT coefficients

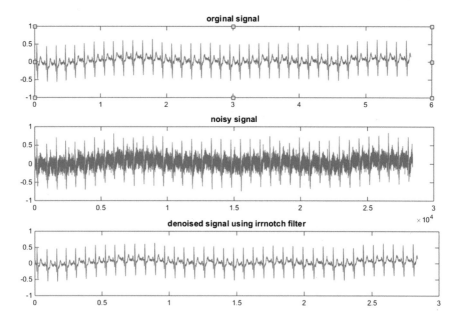

Fig. 8.16 Denoised signals with hard threshold signal to DWT coefficients

8.8 Conclusion

In this paper, the significance of basic filters for removal of noise is proved. This paper addressed denoising techniques with time and frequency domain filtering techniques. The results are compared with various performance measures like PSNR, SNR, MSE. Denoising worked well in frequency domain instead of time domain for noise removal. It can consider as basic review paper for further extension of denoising algorithms.

References

1. Agnate PM et al (1999) ECG noise filtering using wavelets with soft thresholding methods. IEEE Comput Cardiol (Germany)
2. Chouakri SA et al (2000) ECG signal smoothening based on combining wavelet denoising signalling levels. Asian J Inf Technol 5(6):666–677
3. Donoho DL (1995) De-noising by soft tresholding. IEEE Trans Inf Theor 41(3):613–627
4. Bui TD, Chen G (1998) Translation invariant denoising using multi wavelets. IEEE Trans Signal Process 46(12):3414–3420
5. Wang Z et al (2016) Adaptive Fourier decomposition based ECG denoising. Comput Biol Med 77:195–205
6. https://www.physionet.org/physiobank/database/html/mitdbdir/mitdbdir.html

7. Nguyen P et al (2016) Adaptive ECG denoising using genetic algorithm-based thresholding and ensemble empirical mode decomposition. Inf Sci 373(10):499–511
8. Xiong P et al (2016) A stacked contractive denoising auto-encoder for ECG signal denoising. Inst Phys Eng Med Physiol Meas 37(12)
9. Das MK (2013) Analysis of ECG signal denoising method based on S-transform. IRBM 34(6):362–370
10. Gupta R et al (2010) Development of an embedded system and MATLAB-based GUI for online acquisition and analysis of ECG signal. Measurement (Elsevier) 43(9):1119–1126
11. Antoniou et al (2016) Digital signal processing, 2nd edn. Mcgraw-Hill, ISBN: 0071846034
12. Chui (1992) An introduction to wavelets. Academic Press, London
13. Bindu CH et al (2012) Performance analysis of multi source fused medical images using multiresolution transforms. IJACSA 3(10):54–62. ISSN: 2156-5570 (Online)

Chapter 9
Design and Implementation of Modified Sparse K-Means Clustering Method for Gene Selection of T2DM

Kumba Vijayalakshmi and Prof M. Padmavathamma

Abstract Clustering is a trouble-free procedure for extracting clusters from large databases. Data clustering is the task of segregating and classifying data elements on the basis of some aspect of resemblance between the elements in the group. If such data encountered with many features, we may face many problems for analysis, for example, "genomic" data in bioinformatics. Nowadays, substantial advancement was achieved in the identification of risk genes involved with Type II Diabetes Mellitus (T2DM). In this paper, we are designing and implementing a modified sparse K-means clustering method for gene selection by collecting microarray data and grouped into clusters to analyze which is the most relevant gene for susceptibility of T2DM. The identification of most relevant genes of the disease is based on expression levels by genome-wide analysis and to make optimal decision results.

Keywords Microarray data · T2DM · K-means clustering
Sparse regularization method · Modified sparse K-means clustering · Gene selection · TCF7L2

9.1 Introduction

Diabetes mellitus is a type of permanent chronic metabolic [1, 2] disorder which arises either due to limited production of insulin in pancreas or when the released insulin is not absorbed by the body. Diabetes mellitus is generally categories into three types. Type I Diabetes Mellitus (T1DM) is consequences of the pancreas malfunction for secretion of insulin. This is common in young age people and children. This accounts to around 5–10% of the diabetes patients. Type II Diabetes Mellitus (T2DM) is developed when the body cells are unable to fully react to insulin. T2DM affects around 90–95% of individuals [3] across the world. It is also considered as adult onset. The main reason for this due to unnecessary body weight, not enough exercises for body, etc. Unlike type-I, it does not depend on the insulin content in the body. It is generally caused due to limited or lower insulin production in the body. Gestational diabetes occurs during pregnancy without a medical history or heredity of diabetes.

© The Author(s) 2019   97
Ch. Satyanarayana et al., *Computational Intelligence and Big Data Analytics*,
SpringerBriefs in Forensic and Medical Bioinformatics,
https://doi.org/10.1007/978-981-13-0544-3_9

Fig. 9.1 Microarray gene
expression experiment

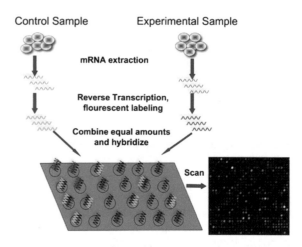

Once the gestational diabetes is developed in women, there are chances of gaining T2DM after post-pregnancy.

When it comes to diabetes, there are few controllable risk variables, including bad lifestyle, lack of exercise, and obesity. The risk factors for T2DM are hypertension, lipids, body mass index (BMI), smoking; low physical activity, unhealthy food habits, heredity, and few particular genes are also highly related as risk factors for T2DM. Until today, genome-wide association studies (GWAS) have recognized some variation within and about 42 genes that are related with vulnerability to T2DM along with general symptoms.

Microarray analysis was conducted to classify the genes [4] by analyzing expression levels of multiple genes at once. Generally, the microarray is a piece of glass slide which is used to mount DNA particles in a systematic way at specific areas known as features or spots. A microarray experiment [5] is composed of a huge number of spots. Every spot will consist of several million copies of indistinguishable DNA particles that particularly relates to a gene.

The large datasets of information from numerous biological experiments are scrutinized and investigated by properly designed genomics. Gene expression analysis is one type of experiment that examines the level of expression of several hundreds of genes, under a specific criterion, which are carried out simultaneously. It is possible with microarray technology where the magnitude of resulting data from each experiment is quite massive. This technology as shown diagrammatically in Fig. 9.1 has turned out to be one of the indispensable approaches for biologists to observe and study expression levels of numerous genes across all genome for the organism in question.

In order to envision the result of data analysis, the gene expression values from microarray experiments are illustrated as heat maps. Examination of gene expression information can be grouped into two types, which are known as supervised and unsupervised learning.

Supervised learning: The annotation and description of gene or sample are used. The gene clusters are created to recognize the patterns of those clusters.

Unsupervised learning: The annotation and description of gene or sample are not used to analyze the information or recognition of patterns. Gene with identical pattern is reluctantly grouped into a cluster.

For making meaningful and valid biological inferences, annotation information must be analyzed using data mining techniques. The genes with identical expression profiles at various experimental conditions are generally considered as a single cluster and used for comparative analysis.

9.2 Importance of Genetic Research in Human Health

The human body is composed of numerous trillions of cells [4]. Every cell consists of a central core with chromosomes, known as nucleus. The chromosomes contain deoxyribonucleic acid, or simply known as DNA. It is comprised of fragments and strands of genes which depict the traits and characteristics of an individual. Each individual has a huge number of genes, which plays a vital part in general well being of a person. Thus, the genetic research assumes an imperative part in identification, inhibition, and additionally treatment of hereditary ailments which are resulted from genetic disorders and gene transformations.

Genetic and biomedical research [4] helps out to discover diseases, health problems that are prone to genetic factors and also to assess an individual for the risk of a particular disease or ailment. Being a carrier of a specific gene is not necessarily associated with getting infected with any disease because majority of health-related issues are triggered by a grouping of quite a lot of factors. On the other hand, individuals who tend to carry certain genes are generally advised to take preventive actions. Consult an expert after a positive genetic test to get risks and advantages of specific preventive mechanisms. Most genetic disorders cannot be cured. The genetic research also involves in the development of treatments by medications or genetic modifications to these genetic abnormalities and mutations which is useful for many people to avoid potentially life-threatening diseases, to restore their health.

9.3 Dataset Description

"T2D-Db" is a widespread database which presents in sequence about the feature concerned with the pathogenesis of T2DM and well structured to assist the researcher's group across the globe. Its external link is http://t2ddb.ibab.ac.in/hom e.html. It encompasses pathophysiology related to T2DM along risk factors for the disease with patient's data. NCBI Entrez gene data also accessible as outward links from this Candidate genes transcript sequences is also available in T2D-Db from

Unigene (NCBI). For experiment analysis 1431 genes Microarray Expression data collected from 28 patients.

9.4   Implementation of Existing K-Means Clustering Algorithm

K-means clustering [6, 7] is the mechanism to characterize a given set of information through a specific number of clusters with a static priori. The principle concept is to assign a centroid for each cluster. The position of the centroid must be chosen carefully, as they tend to have several effects on the results.

K-means clustering algorithm usually works according to the following steps:

Existing K-means Clustering algorithm:

Step 1:   Input data points (X) and set of centers (V), i.e., let set of data points be denoted as $X = \{x_1, x_2, x_3, \ldots\ldots, x_n\}$ and the set of centers be denoted as $V = \{v_1, v_2, \ldots\ldots, v_c\}$.

Step 2:   A 'c' cluster centers are selected at random.

Step 3:   Distance between each cluster centers and data points is calculated.

Step 4:   The data point is assigned to the cluster center whose distance from the cluster center is least of the involving cluster centers.

Step 5:   The fresh cluster center is recalculated using the equation,

$$v_i = (1/c_i) \sum_{j=1}^{c_i} x_i$$

where "c_i" denotes the number of data points in ith cluster.

Step 6:   The distance between new obtained cluster centers and data point is recalculated.

Step 7:   If no. of more data point is reassigned, then halt the process, or else, go to step 4.

For allocation of objects into clusters, K-means uses the squared Euclidean distance parameter. It is believed that the data must appropriately present on the identical scale to make use of such distances. While execution, it attempts to estimate the least value of the mean square error (MSE). The process is executed several times with varying preliminary values and we get dissimilar clusters with disparate cluster vectors. Here, we required to make out the number of clusters you look for to find the optimal solution. While implementing this clustering technique, the number of clusters is fixed, and clustering error is reported which is caused due to distance error. This error is known as *clustering error*. This is estimated by computing sum of squared distance between cluster center and data point.

9.5 Implementation of Proposed Modified Sparse K-Means Clustering Algorithm

In order to improve the clustering performance, for reducing cluster error we incorporated new optimization approach called sparse regularization theory for K-means clustering algorithm and implemented on above-mentioned microarray dataset. To reduce the cluster error obtained by K-means clustering, we are constructing cluster membership matrix by using optimization equation

$$\min_{U \in \{0, 1\}^{n \times C}} \quad \max_{\{c_k(.)\} \in \mathcal{H}\}_{k=1}^{C}} \sum_{k=1}^{C} \sum_{i=1}^{n} \frac{U_{i,k}}{n} |c_k(.) - k(x_i, .)|_{\mathcal{H}_k}^2$$

U is the cluster membership matrix satisfying $U_{i,k} = 1$, if data point g_i belongs to cluster k and 0 otherwise, $\{c_k(.)\}_{k=1}^{C}$ denote cluster center, factor $\frac{1}{n}$ is used for time span.

To get sparse optimal weights, we applied L_1 *regularization* method. It is one-type method of method in sparse regularization method [8]. Sparsity regularization methods focus on selecting the input variables that best describe the output. This method utilizes the assumption that the output variable to be learned can be described by a reduced number of variables in the input space. It is interpretability, high-dimensional learning method used where dimensionality of data may be higher than the number of observations, and reduction of computational complexity. The enhanced sparse K-means clustering algorithm is as follows:

Step 1: Initialize the Gene Id as input that has microarray expression value points to

G, P denotes Patient Id and set of centers V

Let set of data points be denoted as $\mathcal{U} = \{g_1, g_2, g_3, \ldots, g_n\}$ where $n = 1$ to 1431; $P = \{p_1, p_2, p_3, \ldots, p_m,\}$ where $m = 0$ to 28 and the set of centers $V = \{v_1, v_2, \ldots, v_n\}$

Step 2: For $i = 1$ to n; for $j = 1$ to m;

Step 2.1: $V[i] = $ Compute "C" cluster centers randomly on $G[i]$.

Step 2.2: Calculate the distance between each data point $G[i]$ and cluster centers $V[i]$.

Step 2.3: Formulate an optimization form based on the Clustering Error, and this optimization form can be represented as

$$\min_{U \in \{0, 1\}^{n \times C}} \quad \max_{\{c_k(.)\} \in \mathcal{H}\}_{k=1}^{C}} \sum_{k=1}^{C} \sum_{i=1}^{n} \frac{U_{i,k}}{n} |c_k(.) - k(x_i, .)|_{\mathcal{H}_k}^2$$

U is the cluster membership matrix satisfying $U_{i,j} = 1$, if data point g_i belongs to cluster k and 0 otherwise, $\{c_k(.)\}_{k=1}^{C}$ denote cluster center, factor $\frac{1}{n}$ is used for time span.

Step 2.4: If $C < V[i]$, then

Assign the $G[i]$ data point to the $C[k]$ cluster center.

Step 3: Recalculate the new cluster center using:

$$v_i = (1/c_i) \sum_{j=1}^{c_i} x_i$$

where "c_i" represents the number of data points in ith cluster.

Step 4: Apply L_1 *regularization* to obtain the sparse optimal weights.

Step 5: Recalculate the distance between each $G[i]$ data point and new obtained $V[i]$ cluster centers and check the optimal solution is obtained or not.

Step 6: If no data point was reassigned then stop, otherwise repeat from step 2.2.

9.6 Results and Discussion

By executing above two algorithms independently on microarray expression data base in different criterions like execution time, total sum of distances, entropy values. The obtained results are tabulated as shown in Table 9.1 (Figs. 9.2 and 9.3).

9.6.1 Cluster Error Analysis

Performance of each clustering vector is evaluated by measuring the clustering error for each cluster vector. In this work, we have considered number of clusters varied from 2 to 30 clusters and mean square error (MSE) is computed during each clustering analysis. As the numbers of clusters are increasing, clustering error also increases in both the scenarios but proposed approach obtains better performance by reducing the cluster error for increasing number of clusters. Clustering error is computed in terms of MSE which is tabulated in Table 9.2.

Clustering error is computed in terms of MSE as depicted in Fig. 9.4.

9.6.2 Selection of More Appropriate Gene from Cluster Vectors

From the above results, we come to know that proposed modified sparse K-means clustering has good performance. Consequently, we executed the algorithm number of times and tabulated number of clusters as well as cluster vector obtained in Table 9.3.

Table 9.1   Results obtained with various criterions

Execution time comparison

No. of records	100	200	300	400	500	600	700	800	900	1000
K-Means (ms)	93	147	282	331	410	523	598	677	739	820
Sparse K-Means (ms)	85	140	263	321	405	440	497	510	659	766

Entropy-based comparison

No. of records	100	200	300	400	500	600	700	800	900	1000
K-Means	0.29	0.33	0.41	0.51	0.59	0.63	0.68	0.71	0.78	0.81
Sparse K-Means	0.32	0.37	0.49	0.58	0.61	0.67	0.71	0.77	0.82	0.91

Best total sum of distances comparison

No. of records	100	200	300	400	500	600	700	800	900	1000
K-Means	369.21	371.89	402.55	421.39	438.1	452.88	459.71	460.33	470.81	483.11
Sparse K-Means	274.1	296.38	310.77	325.17	333.72	346.31	360.78	390.17	416.88	439.75

Table 9.2   Cluster error is computed in terms of MSE

Number of clusters	2	5	8	10	15	20	25	30
MSE-K-Means	5.58	6.39	6.88	9.34	11.06	11.74	12.37	13.08
MSE-Sparse K-Means	4.31	4.81	5.06	7.18	8.11	8.44	9.30	10.13

In the above given table (Table 9.3) with size of 10 iterations, we formulate the clustering vector where it shows that each cluster consists of ***TCF7L2*** gene. It indicates that it is most relevant suspicious gene which affects insulin secretion and glucose production that leads to T2DM.

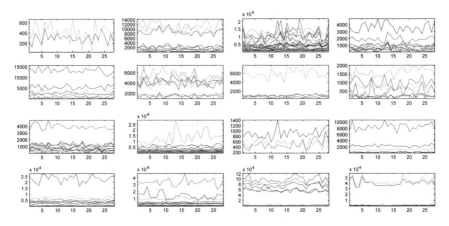

Fig. 9.2 K-means clustering of expression profile values

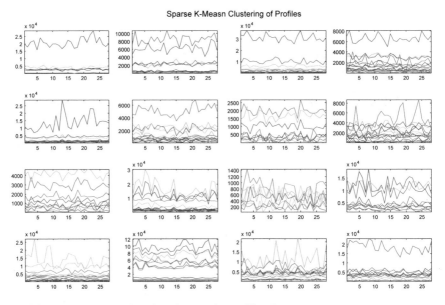

Fig. 9.3 Sparse K-means clustering of expression profile values

9.7 Conclusion

In this paper, we integrated sparse regularization theory to existing K-means algorithm to reduce cluster error and also implemented new modified sparse K-means clustering algorithm on microarray dataset. We analyzed the obtained cluster vectors; confirm that TCF7L2 gene is most susceptible, strongest well-known genetic risk factor that affects T2DM.

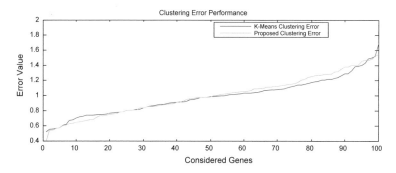

Fig. 9.4 Cluster error performance in terms of MSE

Table 9.3 Cluster vectors computed with various center sizes

Centers	Size of clusters	Clustering vectors
2	14,1	1{ADRB3, APOA2, BDNF, C4A, CACNA1D, ACNB3, CAMK2G, HFE, HK2, HNF4A, IGF2, KCNA3, KCNJ10, *TCF7L2*};2{HIF1A};
3	13,1,1	1{ADRB3, APOA2, BDNF, C4A, CACNA1D, CACNB3, CAMK2, HFE, HK2 HNF4A, IGF2, KCNA3, KCNJ10};2{HIF1A};*3{TCF7L2};*
4	10,1,1,3	1{ADRB3, APOA2, C4A, CACNA1D, CACNB3, HFE, HNF4A, IGF2, KCNA3 KCNJ10};*2{TCF7L2};*2{HIF1A};4{BDNF, CAMK2G, HK2,};
5	3,3,7,1,1	1{C4A, CACNA1D, CACNB3};2{BDNF,CAMK2G, HK2}; 3{ADRB3, APOA2, HFE, HNF4A, IGF2, KCNA3, KCNJ10};*4{TCF7L2};*5{HIF1A};
6	7,2,3,1,1,1	1{ADRB3, APOA2, HFE, HNF4A, IGF2, KCNA3, KCNJ10};2{BDNF, CAMK2G}; 3{C4A, CACNA1D, CACNB3}; 4{HIF1A}; 5{HK2};*6{TCF7L2};*
7	1,3,7,1,1,1,1	*1{TCF7L2};*2{C4A, CACNA1D, CACNB3} 3{ADRB3, APOA2, HFE, HNF4A, IGF2,KCNA3,KCNJ10} 4{HK2};5{HIF1A};6{CAMK2G}7{BDNF}
8	2,1,1,1,1,5,2,2	1{HFE,KCNJ10};2{HIF1A};3{HK2};*4{TCF7L2};*5{CACNB3}; 6{ADRB3,APOA2,HNF4A,IGF2,KCNA3}; 7{CAMK2G}; 8{C4A,CACNA1D};
9	5,1,1,1,2,1,2,1,1	1{ADRB3,APOA2,HNF4A,IGF2,KCNA3};2{HIF4A};3{HK2}; 4{CACN1D}; 5{BDNF,CAMK2G}; *6{TCF7L2};* 7{HFE,KCNJ10}; 8{CACNB3}; 9{C4A};
10	2,1,1,1,2,4,1,1,1,1	1{C4A,CACNA1D};2{HNF4A};3{HK2};4{CACNB3}; 5{BDNF,CAMK2G};6{ADRB3,APOA2,IGF2,KCNA3}; 7{KCNJ10};8{HFE};*9{TCF7L2};*10{HIF4A};

References

1. American Diabetes Association (2007) Standards of medical care in Diabetes—2007. Diab Care 30(1):S4–S41
2. https://en.wikipedia.org/wiki/Diabetes_mellitus
3. http://apps.who.int/iris/bitstream/10665/204871/1/9789241565257_eng.pdf Global Report on Diabetes by WHO
4. Horenstein RB, Alan R Genetics of Diabetes. Shuldiner University of Maryland School of Medicine, Division of Endocrinology, Diabetes and Nutrition, 660 West Redwood Street, Room 494, Baltimore, MD 21201, USA
5. http://www.mrc-lmb.cam.ac.uk/genomes/madanm/microarray/chapter-final.pdf
6. Kanungo T, Mount DM, Netanyahu NS, Piatko CD, Silverman R, Wu AY (2002) An efficient k-means clustering algorithm: analysis and implementation
7. Dunham MH (2003) DataMining; Introductory and advanced topics. Pearson Edition
8. Sparsity based regularization, statistical learning theory and applications, March 8th, 2010 by Lorenzo Rosasco Scribe: Ioannis Gkioulekas

Chapter 10
Identifying Driver Potential in Passenger Genes Using Chemical Properties of Mutated and Surrounding Amino Acids

Jayanta Kumar Das, Richa Singh, Pabitra Pal Choudhury and Bidyut Roy

Abstract One of the major challenging tasks today is to understand uncontrolled growth of tumor. Generally, two types of genes and mutations are observed in tumor cells. The driver mutations, within genes, confer a selective growth advantage and are responsible for causing the cancer. The passenger mutations are observed in those genes that, generally, do not provide growth advantage of cells in tumor. In tumor genome, more frequently mutated genes are considered as driver, whereas less frequent mutations are known as passenger genes. The prime aim of the present article is to identify the set of passenger genes that may have driver potential. The current analysis completely deals with the amino acid sequence and embedded chemical properties of amino acids present in both driver and passenger proteins. We picked up mutated and surrounding 21 amino acids, i.e. one mutated non-synonymous amino acid in the middle and 10 amino acids on the both sides of the mutated amino acid, in driver proteins and compared the presence of this length of amino acids having the same mutated amino acid, in passenger protein sequence. In this comparison of pairwise alignment, we generated similarity score between driver and passenger proteins. Based on the similarity index (i.e., alignment score) above the median value, we considered a set of passenger genes as having driver potential. Some of these passenger genes also possess reported biological functions and are found in the pathways of cancer development. So, these passenger genes or proteins, with diver potential, may play crucial role for cancer development.

Keywords Somatic mutations · Driver genes · Passenger genes · Chemical properties of amino acids · Similarity index · Driver potential

10.1 Introduction

Recent observations have demonstrated that the mutations are took place at multiple points in the genome of cancer tissue [1–3]. Generally, mutated genes/proteins are divided into two groups such as driver and passenger genes/proteins on the basis of frequency of mutations and the role they play in carcinogenesis. Accordingly, genes/proteins are designated as driver and passenger in cancer cells. The driver

© The Author(s) 2019
Ch. Satyanarayana et al., *Computational Intelligence and Big Data Analytics*,
SpringerBriefs in Forensic and Medical Bioinformatics,
https://doi.org/10.1007/978-981-13-0544-3_10

genes, providing selective advantage for tumor growth, are more effective in causing cancer phenotype and the passenger mutations are less important functionally in causing cancer phenotype. But, the abundance of passenger genes is much more than driver genes and passenger mutations are considered as background mutations in tumor cells [4]. Some of the mutations in passenger genes can have potentially deleterious effects like driver mutation on the growth of cancer cells [5, 6]. Researchers are studying by different methods to find the passenger and driver genes in different cancers [7–10]. It is interestingly noted that the mutation (driver or passenger) in DNA might cause change of amino acid which may be synonymous or non-synonymous. A non-synonymous mutation alters the amino acid in the sequence of protein. On the other hand, synonymous mutation does not alter amino acid sequences although there are alterations in gene sequence. Therefore, it is informative and useful to know whether non-synonymous mutations and the context of surrounding amino acids observed in driver protein are also present in passenger proteins in a cancer type and implication of this block of amino acids present in passenger genes. Computationally, finding of such passenger genes or mutations is quite difficult due to the huge number of passenger genes having various kinds of mutations. So, we have chosen a computationally intelligent and feasible method that has some biological significance.

In this manuscript, mutation data of known driver and passenger genes from oral cancer tissues was taken but protein sequences of driver and passenger genes were used for analysis. We compared mutated and surrounding 21 amino acids in these driver proteins with that present in any passenger protein, having the same mutated amino acid in the middle of 21 amino acids, to get a score for passenger proteins. These scores were analyzed to explore the potential role of these passenger genes/proteins as a driver.

10.2   Materials and Methods

10.2.1   Dataset Specification

In this study, we used mutation data in 15 driver and 2161 passenger genes which were generated from whole exome sequencing of 11 different oral cancer tissues in our laboratory [unpublished] and corresponding protein sequences were taken from NCBI database [11]. Driver and passenger gene mutation data were obtained from 7 and 11 cancer samples, respectively (Table 10.1).

Table 10.1 Dataset specification of mutated driver and passenger genes from oral cancer

Gene category	Number of genes	Number of cancer samples containing these genes	Number of chromosomes containing these genes
Driver gene	15	7	3
Passenger gene	2161	11	23

10.2.2 Computational Methodology

There are twenty different amino acids that make up proteins. Each amino acid can be assigned by one-letter codes (upper case) (Table 10.2) [12–14].

For block-specific comparison between the driver and passenger proteins, we categorized all amino acids into eight chemical groups (Table 10.3) [15, 16]. There are various methods that are used for homolog detection [17–20]. Recently, a method has been discussed [15, 16] and found to be useful for the homolog detection and identification of functional zone [20]. So, we rely on that method which may be expedient for the prediction of driver potential in passenger genes. We selected a block of 21 amino acids including the locus of point mutation at middle and 10 amino acids at both left and right sides of the locus of mutation point.

Similarity Index (*SI*): The similarity index (*SI*) for two blocks ($A_1, A_2 \ldots A_l$ and $B_1, B_2 \ldots B_l$) of amino acid sequences of same length (l) and their chemical groups ($G_k; k \in \{1, 2, 3 \cdots 8\}$) is calculated as follows.

$$SI = \left\lceil \frac{\sum_{i=1}^{l} S(A_i, B_i)}{l} \times 100 \right\rceil \qquad (10.1)$$

Table 10.2 Twenty standard amino acids and their one letter code

Amino acids	One letter code	Amino acids	One letter code
Aspartic acid	D	Valine	V
Glutamic acid	E	Alanine	A
Arginine	R	Glycine	G
Histidine	H	Proline	P
Lysine	K	Methionine	M
Tyrosine	Y	Cysteine	C
Phenylalanine	F	Serine	S
Tryptophan	W	Threonine	T
Isoleucine	I	Glutamine	Q
Leucine	L	Asparagine	N

Table 10.3 Twenty standard amino acids with one letter code and their chemical properties divided into eight groups [15]

Chemical properties	Acidic	Basic	Aromatic	Aliphatic	Cyclic	Sulfur containing	Hydroxyl containing	Acidic amide
Amino acids	D, E	R, H, K	Y, F, W	I, L, V, A, G	P	M, C	S, T	Q, N
Group number	1	2	3	4	5	6	7	8

where

$$S(A_i, B_i) = \begin{cases} 1, & \text{if } A_i, B_i \in G_k \\ 0, & \text{otherwise} \end{cases}$$

For an example, let there be two amino acid sequences GDLVCRN of block length 7 where there is a mutation on V (bold) and another amino acid sequences LRIVMQE of same length block. For alignment of these two blocks, pairwise four columns $((G, L), (L, I), (V, V), (C, M))$ are observed from same chemical groups (Table 10.3), so the similarity index (SI) between these two blocks is $(\frac{4}{7} \times 100) = 58$ as calculated by Eq. (10.1).

For the selection of a threshold value of similarity scores to be considered, we have calculated median (SI_{Th}) for the set of similarity indexes obtained by block comparison. Above the median values, the identified passenger genes, were considered as having potential driver property.

10.3 Results and Discussions

10.3.1 Mutations in Both the Driver and Passenger Genes

We found 20 different mutations in 15 driver genes and 158 different mutations in 2161 passenger genes (Table 10.4). It is also interestingly noted that all the mutations that are present in driver genes are also present in passenger genes, although amino acid contexts are different. Therefore, these 20 mutations are unique in driver genes, but there are 138 (158 − 20) mutations that are unique in passenger genes. We focused on 20 mutations common to driver and passenger genes (highlighted in yellow color) and corresponding mutated amino acids were represented by one letter code. A mutation in amino acid (or change of amino acid) $A \rightarrow V$ was due to genetic mutation from C to T in nucleotide sequence of the gene and it was non-synonymous mutation

Table 10.4 Mutations in driver and passenger proteins based on amino acid change

Mutations in Driver protein	Mutations in Passenger proteins							
A→V	A→D	D→Y	G→V	K→M	M→T	Q→P	S→I	V→G
C→S	A→E	E→A	G→W	K→N	M→V	Q→R	S→L	V→I
E→K	A→G	E→D	G→X	K→Q	N→D	Q→X	S→N	V→L
E→Q	A→P	E→G	H→D	K→R	N→H	R→C	S→P	V→M
E→V	A→S	E→K	H→L	K→T	N→I	R→G	S→R	W→C
F→I	A→T	E→Q	H→N	K→X	N→K	R→H	S→T	W→G
G→E	A→V	E→V	H→P	L→F	N→S	R→I	S→W	W→L
G→S	C→F	E→X	H→Q	L→H	N→T	R→K	S→X	W→R
G→V	C→G	F→C	H→R	L→I	N→Y	R→L	S→Y	W→S
I→V	C→R	F→I	H→Y	L→M	P→A	R→M	T→A	W→X
L→S	C→S	F→L	I→F	L→P	P→H	R→P	T→I	X→G
L→V	C→W	F→S	I→L	L→Q	P→L	R→Q	T→K	X→W
M→I	C→X	F→V	I→M	L→R	P→Q	R→S	T→M	Y→C
N→S	C→Y	F→Y	I→N	L→S	P→R	R→T	T→N	Y→D
R→H	D→A	G→A	I→R	L→V	P→S	R→W	T→P	Y→F
R→X	D→E	G→C	I→S	L→W	P→T	R→X	T→R	Y→H
S→X	D→G	G→D	I→T	L→X	Q→E	S→A	T→S	Y→S
T→N	D→H	G→E	I→V	M→I	Q→H	S→C	V→A	Y→X
V→M	D→N	G→R	K→E	M→L	Q→K	S→F	V→D	
Y→X	D→V	G→S	K→I	M→R	Q→L	S→G	V→F	

The highlighted yellow color mutations in passenger genes are also present in driver genes

from A (Alanine) to V (Valine) in the corresponding protein sequence. Here, we worked on the mutated non-synonymous amino acids rather than mutated nucleotides because mutations in nucleotide may not change the amino acid.

Here, change in chemical group or side chain of amino acids (Table 10.3) (e.g., $F \rightarrow I$ means change in chemical group number from 3 to 4) for all the mutations was considered (Table 10.5). If we look into the change in chemical group in the amino acid due to mutation, then we find few mutations where chemical group of amino acids did not change due to mutation. They belonged to chemical groups 4 and 1 (Table 10.5) (i.e., after mutations amino acids remained as aliphatic, $4 \rightarrow 4$ or acidic group, $1 \rightarrow 1$).

Maximum number of mutations was found in TP53 driver gene (5 mutations) which is also observed in other studies on oral/head and neck cancer [21]. The mutations $A \rightarrow V$ and $E \rightarrow K$ were found on two different driver genes but remaining mutations were found in only one gene (Table 10.6). But, $A \rightarrow V$ and $E \rightarrow K$ mutations are found in 80 and 52 passenger genes, respectively. The mutations $A \rightarrow T$ were observed in 91 passenger genes but were absent in driver gene.

Table 10.5 Mutation in driver and passenger gens with regard to the change of chemical group

Mutations in Driver proteins	Mutations in Passenger proteins							
4→4	4→1	1→3	4→4	2→6	6→7	8→5	7→4	4→4
6→7	4→1	1→4	4→3	2→8	6→4	8→2	7→4	4→4
1→2	4→4	1→1	4→X	2→8	8→1	8→X	7→8	4→4
1→8	4→5	1→4	2→1	2→2	8→2	2→6	7→5	4→6
1→4	4→7	1→2	2→4	2→7	8→4	2→4	7→2	3→6
3→4	4→7	1→8	2→8	2→X	8→2	2→2	7→7	3→4
4→1	4→4	1→4	2→5	4→3	8→7	2→4	7→3	3→4
4→7	6→3	1→X	2→8	4→2	8→7	2→2	7→X	3→2
4→4	6→4	3→6	2→2	4→4	8→3	2→4	7→3	3→7
4→7	6→2	3→4	2→3	4→6	5→4	2→6	7→4	3→X
4→4	6→7	3→4	4→3	4→5	5→2	2→5	7→4	X→4
6→4	6→3	3→7	4→4	4→8	5→4	2→8	7→2	X→3
8→7	6→X	3→4	4→6	4→2	5→8	2→7	7→6	3→6
2→2	6→3	3→3	4→8	4→7	5→2	2→7	7→8	3→1
2→X	1→4	4→4	4→2	4→4	5→7	2→3	7→5	3→3
7→X	1→1	4→6	4→7	4→3	5→7	2→X	7→2	3→2
7→8	1→4	4→1	4→7	4→X	8→1	7→4	7→7	3→7
4→6	1→2	4→1	4→4	6→4	8→2	7→6	4→4	3→X
3→X	1→8	4→2	2→1	6→4	8→2	7→3	4→1	
	1→4	4→7	2→4	6→2	8→4	7→4	4→3	

The highlighted yellow color mutations in passenger genes are also present in driver genes

From this study, it is observed that mutations in driver genes are also present in passenger genes but the context of amino acid sequence around mutations is different in both the genes. Here, we found that this context of amino acid sequence around mutation in driver genes is also present to some extent in passenger genes around the same mutations. So, we predict that some of the passenger genes could have potential to function as driver genes. Like this study, some other reports also predicted driver potential of some passenger genes using different methodology [22–24].

10.3.2 Block-Specific Comparison Driver Versus Passenger Protein

Using Eq. (10.1), we calculated similarity index for every mutation by comparing the driver gene block of 21 amino acids with the respective passenger gene block of 21 amino acids around the same mutation point. The similarity index (SI) between

Table 10.6 Frequency of mutations observed in different driver and passenger genes

Driver Mutations		Passenger Mutations											
Mutation	Gene count	Mutation	Gene count	Mutation	Gene count	Mutation	Gene count	Mutation	Gene count	Mutation	Gene count	Mutation	Gene count
A→V	2	A→D	17	F→V	2	K→T	8	Q→E	16	S→Y	2
C→S	1	A→E	10	F→Y	3	K→X	3	Q→H	22	T→A	62
E→K	2	A→G	12	G→A	12	L→F	37	Q→K	10	T→I	32
E→Q	1	A→P	17	G→C	5	L→H	4	Q→L	4	T→K	5
E→V	1	A→S	46	G→D	14	L→I	8	Q→P	6	T→M	47
F→I	1	A→T	91	G→E	16	L→M	6	Q→R	31	T→N	12
G→E	1	A→V	80	G→R	46	L→P	33	Q→X	9	T→P	3
G→S	1	C→F	11	G→S	36	L→Q	5	R→C	65	T→R	4
G→V	1	C→G	2	G→V	18	L→R	4	R→G	19	T→S	14
I→V	1	C→R	15	G→W	2	L→S	3	R→H	78	V→A	46
L→S	1	C→S	2	G→X	3	L→V	30	R→I	1	V→D	1
L→V	1	C→W	2	H→D	9	L→W	1	R→K	26	V→F	10
M→I	1	C→X	4	H→L	5	L→X	2	R→L	16	V→G	6
N→S	1	C→Y	13	H→N	4	M→I	20	R→M	1	V→I	64
R→H	1	D→A	6	H→P	2	M→L	6	R→P	10	V→L	22
R→X	1	D→E	19	H→Q	9	M→R	2	R→Q	83	V→M	39
S→X	1	D→G	13	H→R	22	M→T	24	R→S	19	W→C	4
T→N	1	D→H	10	H→Y	15	M→V	25	R→T	5	W→G	1
V→M	1	D→N	44	I→F	2	N→D	20	R→W	34	W→L	2
Y→X	1	D→V	6	I→L	13	N→H	3	R→X	9	W→R	9
		D→Y	15	I→M	16	N→I	2	S→A	15	W→S	1
		E→A	9	I→N	2	N→K	26	S→C	13	W→X	5
		E→D	35	I→R	1	N→S	38	S→F	22	X→G	1
		E→G	19	I→S	6	N→T	4	S→G	23	X→W	1
		E→K	52	I→T	23	N→Y	1	S→I	9	Y→C	16
		E→Q	23	I→V	59	P→A	20	S→L	25	Y→D	2
		E→V	9	K→E	32	P→H	7	S→N	34	Y→F	7
		E→X	9	K→I	3	P→L	52	S→P	20	Y→H	10
		F→C	4	K→M	1	P→Q	9	S→R	20	Y→S	1
		F→I	6	K→N	25	P→R	16	S→T	19	Y→X	4
		F→L	21	K→Q	8	P→S	35	S→W	3		
		F→S	6	K→R	31	P→T	17	S→X	2		

The highlighted yellow color mutations in passenger genes are common to driver mutations. gene count: mutation present in number of genes

driver and passenger proteins ranges from 48 (maximum) to 24 (minimum). This suggests that the block-specific similarity index based on the embedded chemical properties of amino acids around mutation point is quite diverse. However, based on the median values (SI_{Th}), we selected a set of passenger genes above the median value and identified the maximum SI_{Th} is 35 and minimum SI_{Th} is 14 for the mutation $A \rightarrow V$ and $G \rightarrow E$, respectively (Table 10.7).

Accordingly, passenger genes having similarity index (SI) more than median value were selected as having driver potential. Among the selected passenger genes,

Table 10.7 Mutation and quantitative information regarding similarity score of passenger genes and biological significance

Mutations in amino acid	Total number of genes	Max. Similarity Index (SI)	Median Value (SI_{Th})	Number of Genes count > SI_{Th}	Number of genes having known functions	Number of genes found in biological pathways
A→V	80	43	29	26	6	1
A→ V	80	48	35	35	8	2
C→ S	2	33	31			
E→ K	52	38	19	20	6	
E→ K	52	33	19	15	3	1
E→ Q	23	38	19	6	3	1
E→ V	9	48	19	4		
F→ I	6	29	19	1	1	
G→ E	16	38	14	6	1	1
G→ S	36	38	24	17	4	3
G→ V	18	29	19	5		
I→ V	59	38	19	20	1	3
L→ S	3	29	24	1	1	1
L→ V	30	48	26.5	15	3	1
M→ I	20	43	24	7	2	1
N→ S	38	38	19	15	1	2
R→ H	78	48	24	28	2	4
R→ X	9	33	19	4	1	
S→ X	2	29	17	1		1
T→ N	12	38	24	5	1	
V→ M	39	43	24	14	2	
Y→ X	4	24	19	1		

Noted that the mutations $A \rightarrow V$ and $E \rightarrow K$ are found in two different driver genes set (RAD23B and ZNF638) and (TP53 and TJP2) respectively

we also identified some genes for which functions are known and some of which are also present in biological pathways involved in carcinogenesis. It is to be noted that, any two passenger genes do not have same mutation except $A \rightarrow V$ (Tables 10.7 and 10.8). The identified passenger genes are called driver potential (Table 10.8) and they may play the crucial role in carcinogenesis.

Table 10.8 Passenger genes with driver potential and their known functions and involvement in biological pathways

Mutations	Driver gene	Passenger gene	Known biological functions	Involvement in biological pathway
A→ V	RAD23B	PXK	Kinase	
		USHBP1	TS	
		HOXB7	Transcription factor	
		PRMT6	TS	
		CDC42BPB	Kinase	
		MYH3		Tight junction protein
		RNF213	Oncogene	
A→ V	ZNF638	EPOR		
		CDC42BPB	Kinase	Jak-stat signaling
		BAK1		Apoptosis
		HOXB7	Transcription factor	
		PRMT6	TS	
		ZNF140	Transcription factor	
		IKBKE	Kinase; TS	
		RNF213	Oncogene	
		TACC2	TS	
		USHBP1	TS	
E→ K	TP53	MBD4	Transcription factor	
		CSMD1	TS	
		NFAT5	Transcription factor	
		MYO3B	Kinase	
		AFF3	Oncogene; transcription factor	
		ATF7IP	Transcription factor	
E→ K	TJP2	SP4	Transcription factor	
		MED16	Transcription factor	
		PIK3CA	Oncogene	Rap1 signaling; foxO signaling
E→ Q	ZFHX3	STK10	Kinase	
		NBN	TS	Cell death
		LILRB1	Cell differentiation marker	
F→ I	SVEP1	EPHA10	Kinase	
G→ E	RPGR	THBS1		pik3-akt pathway
		STK31	Kinase	
G→ S	ZFP36L2	PPP2R3A		pik3-akt signaling
		NACC1	TS	
		PHF24	TS	
		RAPGEF3		Rap1 signaling
		TRIP11	Oncogene; TF	
		SMG1		
		SKAP1	Kinase	Rap1 signaling

(continued)

Table 10.8 (continued)

Mutations	Driver gene	Passenger gene	Known biological functions	Involvement in biological pathway
I→ V	TET2	CTSC		Apoptosis
		GSTP1		Drug resistance
		ITGAE	Cell differentiation marker	Cell adhesion
L→ S	SF3B1	TP73	TF; TS	p53 pathway
L→ V	TAOK2	CEACAM8	Cell differentiation marker	
		CIITA	Oncogene; TF; TS	
		KAT6B	TS	
		SPTA1		Apoptosis
M→ I	WNK1	FGF12	Growth factor	Mapk pathway
		FAM3B	Growth factor	
N→ S	ZNF638	ALCAM	Cell differentiation marker	Cell adhesion
		LRP2		Hedgehog signaling
R→ H	TAOK2	IL17C	Growth factor	
		VCAN		Cell adhesion
		CASZ1	TF	
		ATP6V1B1		mtor pathway
		GORAB		p53 signaling
		PLA2G2E		Ras signaling
R→ X	TP53	CR1	Cell differentiation marker	
S→ X	ZC3H11A	HTR7		Ras signaling pathway
T→ N	SRGAP3	MRO	TS	
		ROCK2	Kinase	
V→ M	ZNF638	KIF13B	TS	
		ADCK3	Kinase	

Passengers genes in bold were found to be potential drivers when compared with two driver genes having $A \rightarrow V$ mutation

10.4 Conclusion

In this study, we have predicted driver potential of some passenger genes by comparing similarity score, deduced from chemical properties of mutated and surrounding amino acids, between driver and passenger genes. Other report [23] also predicted driver potential of some genes using copy number variation and expression of genes in the deleted regions of chromosomes by a pipeline called ProcessDriver. Another report [24] also predicted driver potential of some rarely mutated genes using mutation and expression data by pipeline called LNDriver. The discussed method was to compare a block of 21 amino acids, being mutation point in the middle of the block and 10 amino acids on the both sides of the mutation point, between the driver and passenger genes and get blockwise similarity score. It is very handy and quite intelligent computational tool as the authors have reported previously [15, 20].

References

1. Sjöblom T, Jones S, Wood LD, Parsons DW, Lin J, Barber TD, Mandelker D, Leary RJ, Ptak J, Silliman N et al (2006) The consensus coding sequences of human breast and colorectal cancers. Science 314(5797):268–274
2. Greenman C, Stephens P, Smith R, Dalgliesh GL, Hunter C, Bignell G, Davies H, Teague J, Butler A, Stevens C et al (2007) Patterns of somatic mutation in human cancer genomes. Nature 446(7132):153–158
3. Stratton MR, Campbell PJ, Futreal PA (2009) The cancer genome. Nature 458(7239):719–724
4. Pon JR, Marra MA (2015) Driver and passenger mutations in cancer. Annu Rev Pathol Mech Dis 10:25–50
5. Luo SY, Lam DC (2013) Oncogenic driver mutations in lung cancer. Transl Respir Med 1(1):6
6. Walczak AM, Nicolaisen LE, Plotkin JB, Desai MM (2011) The structure of genealogies in the presence of purifying selection: A "fitness-class coalescent". Genetics 111
7. Fröhling S, Scholl C, Levine RL, Loriaux M, Boggon TJ, Bernard OA, Berger R, Döhner H, Döhner K, Ebert BL et al (2007) Identification of driver and passenger mutations of FLT3 by high-throughput DNA sequence analysis and functional assessment of candidate alleles. Cancer Cell 12(6):501–513
8. Dimitrakopoulos CM, Beerenwinkel N (2017) Computational approaches for the identification of cancer genes and pathways. Wiley Interdisc Rev Syst Biol Med 9(1):e1364
9. Carter H, Chen S, Isik L, Tyekucheva S, Velculescu VE, Kinzler KW, Vogelstein B, Karchin R (2009) Cancer-specific high-throughput annotation of somatic mutations: computational prediction of driver missense mutations. Cancer Res 69(16):6660–6667
10. Nei M, Gojobori T (1986) Simple methods for estimating the numbers of synonymous and nonsynonymous nucleotide substitutions. Mol Biol Evol 3(5):418–426
11. https://www.ncbi.nlm.nih.gov
12. Das JK, Majumder A, Choudhury PP, Mukhopadhyay B (2016) Understanding of genetic code degeneracy and new way of classifying of protein family: a mathematical approach. In: 2016 IEEE 6th international conference on advanced computing (IACC). IEEE, pp 262–267
13. https://teaching.ncl.ac.uk/bms/wiki/index.php
14. Meister A et al (1957) Biochemistry of the amino acids
15. Das JK, Das P, Ray KK, Choudhury PP, Jana SS (2016) Mathematical characterization of protein sequences using patterns as chemical group combinations of amino acids. PloS One 11(12):0167651
16. Das JK, Choudhury PP (2017) Chemical property based sequence characterization of PpcA and its homolog proteins PpcB-E: a mathematical approach. PloS One 12(3):0175031
17. Weathers EA, Paulaitis ME, Woolf TB, Hoh JH (2004) Reduced amino acid alphabet is sufficient to accurately recognize intrinsically disordered protein. FEBS Lett 576(3):348–352
18. Murphy LR, Wallqvist A, Levy RM (2000) Simplified amino acid alphabets for protein fold recognition and implications for folding. Protein Eng 13(3):149–152
19. Li T, Fan K, Wang J, Wang W (2003) Reduction of protein sequence complexity by residue grouping. Protein Eng 16(5):323–330
20. Basak P, Maitra-Majee S, Das JK, Mukherjee A, Dastidar SG, Choudhury PP, Majumder AL (2017) An evolutionary analysis identifies a conserved pentapeptide stretch containing the two essential lysine residues for rice L-myo-inositol 1-phosphate synthase catalytic activity. PloS One 12(9):0185351

21. Van Rechem C, Whetstine JR (2014) Examining the impact of gene variants on histone lysine methylation. Biochim Biophys Acta (BBA) Gene Regul Mech 1839(12):1463–1476
22. Wu H-T, Hajirasouliha I, Raphael BJ (2014) A combinatorial algorithm to identify independent and recurrent copy number aberrations across cancer types. In: 2014 IEEE 4th international conference on computational advances in bio and medical sciences (ICCABS). IEEE, p 1
23. Baur B, Bozdag S (2017) Processdriver: a computational pipeline to identify copy number drivers and associated disrupted biological processes in cancer. Genomics 109(3):233–240
24. Wei P-J, Zhang D, Xia J, Zheng C-H (2016) LNDriver: identifying driver genes by integrating mutation and expression data based on gene-gene interaction network. BMC Bioinform 17(17):467

Chapter 11
Data Mining Efficiency and Scalability for Smarter Internet of Things

M. Mahendra, C. Kishore and Ch. Prathima

Abstract Smarter Internet of Things (SIoT) is defined to transform all areas of our lives. The amount of entities linked to smarter IoT is likely to achieve 50 billion by 2020, which may retort a large amount of priceless data. The information generated from the smarter IoT devices will be utilized to comprehend and organize sophisticated environments of our lives, to enable good decision making, and great level of automation, higher efficiencies, higher production, and accuracy. Information mining and other AI systems would participate a crucial responsibility in making a smarter IoTs, with great challenges. In this work, we analyze the six recognized information mining algorithms for smarter IoT data and also included machine learning and ANN for modeling complex data abstraction. The experimental results on a real smarter IoT dataset illustrate C4.5 and C5.0 which has good reliability and accurateness than the remaining algorithms. ANNs produce high precise results but are classy in computation.

Keywords Smarter Internet of things · Support vector machine classifier
K-nearest neighbors rule · Artificial neural networks · Machine learning

11.1 Introduction

The smarter Internet of things (SIoT) is "a inclusive infrastructure for information society, allowing superior services by connecting (virtual and physical) things based on presented and growing interoperable data and communication technologies" [1]. IoT [2, 3] is established to transform all areas of our lives, social communication, or amusement. Various things are growing day by day to be a part of smarter IoT, and the amount of things linked to IoT is likely to reach 50 billion by 2020. This might create large amount of classy data.

A key goal of IoT is to create our surroundings smarter, by providing the environment the data it requires, through synchronize and historical record feeds, and pertains computations on the data to take smart decisions automatically. The information gathered from SIoT devices will be utilized to realize and organize sophisticated

© The Author(s) 2019
Ch. Satyanarayana et al., *Computational Intelligence and Big Data Analytics*,
SpringerBriefs in Forensic and Medical Bioinformatics,
https://doi.org/10.1007/978-981-13-0544-3_11

surroundings, allowing better good making, superior automation, and higher efficiencies. A major challenge in these configurations is that it takes time to analyzing vast data (i.e., big data) to create good consistent and precise insights and decisions so that SIoT could surpass its promises. Artificial intelligence is one of the best solutions to achieve concealed insights from SIoT data.

Main purpose of this work is to investigate if the predictable information mining algorithms could also work for the SIoT dataset or innovative groups of information mining algorithms will be required. This work present a study on examining the applicability of numerous recognized information mining algorithms to SIoT dataset. In this work, we used six information mining algorithms. They are support vector machine classifier (SVMC), K-nearest neighbors rule (KNNR), naive Bayes theorem (NBT), C4.5, C5.0, ANNs. The main contribution of the paper is the analysis of the performance and effectiveness of six of the recognized information mining algorithms.

11.2   Background Work and Literature Review

Nowadays, information mining responsibilities and challenges increase due to the unprecedented upsurge in the total amount and intricacy of data [4]. Introduction of SIoT model, a totally latest group of obstacles are put into the information mining area [5, 6]. Support vector machine classifier, K-nearest neighbor rule (KNNR), naive Bayes theorem (NBT), C4.5, C5.0, and ANNs are trusted in neuroscientific data mining. Support vector machine classifier has been designed primarily to handle variance of bias trade-off, above-fitting and capability control. Burges and support vector machine classifier explained that exactness depends on a great deal of training data and device capability. Usage of support vector machine classifier is additional prolonged as of classification toward regression and dynamic rating. Support vector machine classifier is an extremely efficient tool to utilize in intricate and loud areas. The computing inefficiency is one of the keys downside of support vector machine classifier; conversely numerous optimizations are performed to lessen its computational cost also to enhance the scalability.

A beginner algorithm is called as KNNR and is one of the easiest accessible classifiers. It is straightforward to comprehend and apply. Because of the straightforwardness of KNNR, numerous issues come up that maximize its performance including the collection of exact distance strategy. In addition, KNNR is used effectively for diverse responsibilities in cordless sensor sites (CSS) and SIoT site for infringement diagnosis, indoor setting systems, and action of identification. In binary classification problems, support vector machine classifier is one of the finest options. However, scalability is obviously a concern. The C4.5 algorithm is one of the finest information mining algorithm, suggested by Ross Quinlan is recognized as the classy algorithm in information mining [7]. Quinlan implemented C5.0 an improvised version of C4.5, which is stated well than C4.5, in memory efficiency, and sustain to enhance, winnowing, and weighting [8].

The method considered by the naive Bayes theorem and ANN systems to resolve the given information mining process is absolutely dissimilar which we have presented priory in this section. The NBT algorithm is an extremely old classifier algorithm. This algorithm uses BT with the self-reliance assumption on the list of characteristics. NBT is a sturdy and straightforward classifier like KNNR. It gives astonishingly exact results where ANN systems are derived from neurosystem of mind. ANNs are really efficient in dealing with information mining responsibilities with high accuracy. Though ANN structured algorithms are also intricate, an intensive quantity of estimation is necessary for solving problems with high accuracy. Future expansion of ANNs will be an innovative models predicated on deep learning principle. ANNs have great learning potential, inputs gigantic amount of information, and produce accurate results that are not likely with traditional Artificial intelligence and information mining algorithms [9]. Though this profound expatiation of computer technology continues to be youth period, deep learning can provide us original insights from SIoT information which is extremely hard from additional information mining algorithms. Especially, regarding SIoT, few works are performed to expand advantages in ANNs.

11.3 Experimental Methodology

We measured six recognized information mining algorithms in the work. The six models similarly embrace ANNs which assemble feedforward multilayer ANNs. The replications are executed by R tool. Especially, for replications ANNs, H2O program used which are available in R tool. For testing, a genuine sensor dataset from the UCI data repository [10] is used. Dataset is obtained by using receptors and accelerometers and is being used to classify real individual activities, automated steering, and human body activities. Before replicating the algorithms, we clean the dataset to be sure that they are perfect for the classifiers. That is clearly a prelude examination, and therefore, we used imperfect dataset. Our testing strategy is depicting in Fig. 11.1.

11.4 Results and Analysis

Here we have presented an investigation on six information mining algorithms as stated in Fig. 11.1. For performance analysis of the algorithms, we have summarized the experimental results form of confusion matrix (CM). By using CM, we have classified total number of instances rightly and wrongly categorized and wrongly classified instances are also to be in CM.

The confusion matrix of six algorithms simulated on a different dataset is shown in Table 11.2, classification accuracy (CA) percentage, and elapsed time.

Considering Table 11.1, we remember that C4.5, C5.0, and ANN models achieved much good performance than SVMC, KNNR, and NBT. With standard accuracy

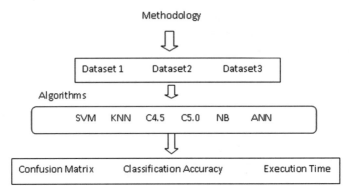

Fig. 11.1 Experimental methodology

of 97.15% acquired by C4.5, it works considerably much improved than 96.61% accuracy of the C5.0 algorithm. Average accuracy of ANNs is 96.19% for all datasets. C4.5 is more accurate among all the six models considered in the classification, following narrowly C5.0.

The dataset is multi-labeled. As a result, SVMC is weak toward multi-labeled information classification when contrast with binary classification in terms of performance which is the finest [7]. SVMC achieves high accuracy than KNNR with 4.09% high. KNNR and distance vector routing algorithm affect the CA of KNNR. The model NBT is not performed well in classification accuracy. Experiments results are highly agreed in the conclusion [7].

11.4.1 Execution Time

NBT algorithm will be the quickest amidst all the six algorithms. Average handling time (AHT) of C4.5, C5.0 is 7.71 and 7.22 mere seconds, respectively. SVMC runs on the good resources of system and has poor dealing out acceleration [11]. KNNR is light process and has low processing times as stated in Table 11.2. ANNs have good system resources. For SIoT, there is a poor classification accuracy which is not considered, but execution time concerns. With instances, NBT is convenient.

11.4.2 Machine Learning Models

In introduction analysis, we assume that ANNs can possess the finest CA among all the replicated models. We noticed that increased classification accuracy would be performed by escalating the eons, neurons, and unknown layers. In ANNs, clas-

Table 11.1 Confusion matrix of (a) SVMC; (b) KNNR; (c) NBT; (d) C4.5; (e) C5.0; (f) ANNs for UCI-HAR dataset

Actual /predicted	Sitting	Sitting down	Standing	Standing up	Walking
a					
Sitting	50,594	3	1	33	1
Sitting down	12	11,523	139	103	50
Standing	2	16	47,127	82	143
Standing up	48	260	267	11,806	34
Walking		106	979	85	42,220
b					
Sitting	46,023	457	3885	258	8
Sitting down	1078	6838	3084	174	653
Standing	306	614	43,852	146	2452
Standing up	1099	2733	5117	1658	1808
Walking	588	2127	98,820	2623	29,232
c					
Sitting	25,366	9	1	9	3
Sitting down	1	5825	48	59	16
Standing	1	4	23,470	44	36
Standing up	14	106	93	5975	23
Walking	5	67	280	55	21,337
d					
Sitting	50,622			9	
Sitting down	5	11,720	18	53	31
Standing		18	47,252	26	74
Standing up	9	55	35	12,264	52
Walking	1	26	73	36	43,254
e					
Sitting	50,616	1	1	13	
Sitting down	13	11,666	52	67	29
Standing		6	47,253	24	87
Standing up	24	90	66	12,189	46
Walking		23	69	24	43,274
f					
Sitting	50,583	6		22	1
Sitting down	9	11,437	31	250	96
Standing		132	47,096	121	237
Standing up	39	173	74	11,951	105
Walking		79	169	71	42,951

Table 11.2 Classification accuracy in percentage and elapsed time in seconds for UCI-HAR dataset

Algorithm	Accuracy (%)	Elapsed time
SVMC	98.76	2351.1
KNNR	98.94	450.6
NB	77.04	0.52
C4.5	99.69	22.65
C5.0	99.62	21.1
ANN	99.03	33,228.1

sification algorithm also is based on its significant variables alteration. ANNs are having a complex framework and need huge amount of system resources, and for that reason, ANN algorithm gets the utmost execution time among all the six models shown in this research.

11.5 Conclusion

The SIoT model conveys new units of information mainly accumulated from the sensor devices. To fully confine, this concealed information from SIoT data is a demanding process in information mining. Fellow research workers dispute a new category of information mining algorithms is need to cope with SIoT data. Inside this research, we evaluate the applications of some developed information mining algorithms including ANNS. With this preliminary evaluation, we intend to perform an in-depth learning on greater and various SIoT dataset in the foreseeable future work.

References

1. Alam Furqan, Mehmood Rashid, Katib Iyad, Albeshri Aiiad (2016) Analysis of eight data mining algorithms for Smarter Internet of Things (IoT), International Workshop on Data Mining in IoT Systems (DaMIS 2016). Procedia Comput Sci 98:437–442
2. Atzori L, Iera A, Morabito G (2010) The Internet of Things: a survey. Comput Netw 54(15):2787–2805
3. Ma H (2011) Internet of Things: objectives and scientific challenges. J Comput Sci Technol 26(6):919–924
4. Cuomo S, Michele PD, Galletti A, Piccialli F (2015) A cultural heritage case study of visitor experiences shared on a social network. In: 2015 10th international conference on P2P, parallel, grid, cloud and internet computing (3PGCIC), Krakow, pp 539–544
5. Chen F, Deng P, Wan J, Zhang D, Vasilakos A, Rong X (2015) Data mining for the internet of things: literature review and challenges. Int J Distrib Sens Netw 2015:1–14
6. Ngai E, Xiu L, Chau D (2009) Application of data mining techniques in customer relationship management: a literature review and classification. Expert Syst Appl 36(2):2592–2602
7. Burges C (1998) A tutorial on support vector machines for pattern recognition. Bell Laboratories and Lucent Technologies

8. Hssina B, Merbouha A, Ezzikouri H, Erritali M (2014) A comparative study of decision tree ID3 and C4.5. Int J Adv Comput Sci Appl Spec Issue Adv Veh Ad Hoc Netw Appl

9. Niu X, Zhu Y, Zhang X. DeepSense: a novel learning mechanism for traffic prediction with taxi GPS traces. In: 2014 IEEE global communications conference (2014)

10. Lichman, M.: UCI machine learning repository (http://archive.ics.uci.edu/ml). University of California, School of Information and Computer Science, Irvine, CA

11. Burbidge R, Buxton B (2001) An introduction to support vector machines for data mining. Computer Science Department, UCL

Chapter 12
FGANN: A Hybrid Approach for Medical Diagnosing

P. Aruna Kumari and Dr. G. Jaya Suma

Abstract Medical diagnostic support systems often deal with a large number of disease measurements and relatively small number of patient records. All these measurements (features) may not be relevant for diagnosing, and some may contain noise due to human or machine errors. These features greatly affect the results of diagnostic systems, and this is significantly high with less number of available patient records. Further, these features will guzzle memory space and time required for diagnosis process. These issues have been addressed in the proposed approach FGANN, which is fuzzy genetic algorithm-based neural network for prediction of disease outcome in the field of health care. In this proposed hybrid approach, the feature space has been modeled using fuzzy approach and then genetic algorithm (GA) has been employed to extract prominent features that show vital impact on diagnosis. These obtained key features are used to train neural network (NN) which in turn used to predict the outcome of the disease for a given patient record. The experiments were carried out on two different types of diseases like diabetics and thyroid by considering standard datasets. In this hybrid approach, not only prediction accuracy, but also the time taken by NN for learning and memory space occupied by patient's information that has been considered as performance measures of the system. The results showed that proposed approach fuzzy logic + GA + NN giving more accurate measures of diagnosis compared to an approach based on NN.

Keywords Genetic algorithm · Fuzzy logic · Artificial neural network
Medical diagnostic system · Feature selection

12.1 Introduction

Differential diagnosis or medical diagnosis is generally perceived as the process of discerning between various diseases which account for patient's health condition based on available data. The intelligent analysis of medical data, particularly for automatic diagnosis, became an unpredictably productive niche in the field of bioinformatics for the rigorous exploitation of soft computing approaches. In this regard,

© The Author(s) 2019
Ch. Satyanarayana et al., *Computational Intelligence and Big Data Analytics*,
SpringerBriefs in Forensic and Medical Bioinformatics,
https://doi.org/10.1007/978-981-13-0544-3_12

while dealing with huge amount of data, the use of soft computing approaches supports medical diagnosis process by making use of the enormous processing capability of computers. When this intelligent system has been fed with various medical data, by an involuntary comparison with data present in medical databases available from high-quality sources produces most probable diagnosis decision. In general, a medical diagnosis has been considered as an endeavor in classifying an individual's medical condition into various classes that permit medical decisions with respect to treatment or predicting next stages of disease [1]. However, diagnosing is not an easy task because symptoms are non-specific. While in the process of decision making, machine learning and soft computing approaches afford support in various areas like health monitoring, medical diagnosis, and different therapies. Although these approaches can improve human decision-making process by possible minimization of physician's error, the final decision must be taken by physician based on computer-aided medical diagnosis system support [2].

Prediction or diagnosis accuracy has usually been the critical goal of prediction or forecasting researches. The accuracy of the prediction model not only depends on the structure of the model and training algorithm employed, nevertheless it also depends on feature space [3]. Feature space is nothing but the set of features (symptoms and measurements) considered for diagnosis. Due to the availability of huge set of features which contains unnecessary, irrelevant, and noisy information, before developing prediction model two major steps need to be performed. The two steps include preprocessing and feature selection. As the medical information is vague, fuzzy theory can be used to efficiently deal with uncertain medical data concepts [4]. Efficient knowledge depiction is one of the vital challenges for the successful construction and following use of medical diagnostic systems in clinical practice [5]. A fuzzy expert system framework has been proposed for diabetic diagnosis by fuzzification of data, generation of fuzzy rules, and defuzzification at the end [6]. In this paper, after preprocessing the features are projected into fuzzy space. Fuzzy membership function has been applied to model linguistic medical information for data to symbol conversion of medical diagnostic support systems.

Another crucial step is feature selection (FS) and is generally used to obtain feature subset from the original set of features by eliminating irrelevant and noisy features which have minimal predictive knowledge [7] and still gives remarkable prediction results [8]. FS has many advantages like better understanding of data, reducing training and implementation time, decreasing storage needs, and raise in predictor performance [9]. In diagnosis process of a disease, the improvement in accuracy and reduction in prediction cost can be achieved by applying FS in association with classification or clustering [10]. Once FS has been chosen, an approach is needed to select prominent features. This selection of directly evaluating all 2^N possible subsets (N number of features) is a NP-hard problem [3]. Therefore, an optimal approach must be applied for FS. In [11], FS approaches have been broadly classified into filter, wrapper, and embedded approaches. In filter approaches, ranks are assigned to features and then best-ranked features are selected as prominent, whereas in wrapper approaches, feature subset is selected based on search algorithm which gives best performance. Embedded approaches find feature subset as a part of training process [8, 12].

However, today's informative field elevating new challenges toward proficient and effective FS approaches due to large volumes of data. In the literature of medical diagnosis, sequential search approaches resulted as successful in several classification applications [13]. A liver tissue has been classified using probabilistic neural network by applying sequential forward selection to obtain reduced feature space [14]. Cardiovascular disease has been predicted by applying a hybrid forward feature selection technique and SVM classifier [15]. Liver lesions have been classified by using neural network classifier by adopting genetic algorithm for FS [13]. Genetic algorithm (GA) has been successfully applied to different applications in classification problems, image processing, and pattern recognition [13, 16]. Because of parallel nature and capability searching complex feature spaces efficiently, GA became popular where traditional FS algorithms fails. Along with neural networks for classification, GA has been proposed to reduce FS for identifying skin tumor [17], while the same methodology has been applied for the classification of microcalcifications [18] and endothelial cells [19].

FS in the context of diagnosis leads to one example of multi-criteria optimization problem. Various criteria to be optimized are classification accuracy, risk, and cost related to classification which in turn relay on set of selected features employed to describe the classification pattern [20]. When compared to traditional approaches for multi-objective optimization problems, evolutionary approaches give better results. This has been motivated toward genetic algorithm for FS. With the reduced feature set, the prediction of a disease greatly affected by the classifier applied. Artificial neural network architecture has been greatly attracted by healthcare research community because of their ability to estimate a random function mapping [21] and in different fields of applications from classification point of view [22]. It became popular in healthcare analytics. A major advantage of this model is that it can handle very complex problems which involve nonlinear relationships between variables [23]. It has been employed in various areas like analysis of cancer cells, analysis of medical signals like ECG and EEG, diagnosis of diabetics, cancer, heart disease and prosthesis design, optimization of hospital cost [24]. High fault tolerance, generalization, and memory capability of neural networks [24] motivated us toward classification of medical data where these make diagnostic system as producing most accurate decisions. The proposed system flow has been depicted in Fig. 12.1.

The rest of the paper is organized as follows: preprocessing of the given medical data along with fuzzification to perform classification task has been described in Sect. 12.2. Genetic algorithm-based feature selection has been discussed in Sect. 12.3. Backpropagation algorithm for classifying the presence of disease has been presented in Sect. 12.4. Section 12.5 presents experimental results and analysis of the proposed system. The paper has been concluded in Sect. 12.6.

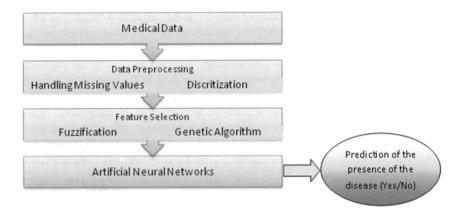

Fig. 12.1 Proposed system flow

12.2 Preprocessing

In this work, the datasets from UCI machine repository have been considered. Before developing a model, the data has to be analyzed and it should be understood to know the structure and relevance of features. The data includes a countable number of missing values for number of features, and some of feature values are continuous and other discrete. An even more noise can be present in the data, which demands the cleaning of data in preparing the dataset for classification analysis [25]. In this hybrid approach, as a part of cleaning, missing values of attributes have been replaced by mean of all the values of the attribute, and based on equal area method, the continuous values have been discretized in this paper.

Fuzzification simply means the process of transforming crisp values into the degree required by the features [6, 26]. Since the medical data generally may generate from hardware, sensor measurements contain ambiguity, vagueness which causes the features to be fuzzy. And this fuzziness can be characterized by using a membership function. A trapezoidal membership function $\mu(a)$ of three parameters (p, q, r, s) has been adopted in this paper, which can be expressed as shown in Eq. (12.1).

$$\mu(a, p, q, r, s) = \begin{cases} 0, & a < p \text{ or } a > s \\ \frac{a-p}{q-p}, & p \le a \le q \\ 1, & q \le a \le r \\ \frac{s-a}{s-r} & r \le a \le s \end{cases} \tag{12.1}$$

where a represents the attribute value to be fuzzified, p is lowest value, s is the highest value of the attribute, q and r are support values for lower and highest values of the attribute in the given data.

12.3 Genetic Algorithm-Based Feature Selection

GA is most popular optimization approach [16]. Darwin's evolutionary mechanics theory has inspired this stochastic search algorithm which consists of fitness, reproduction, mutation, and crossover [26]. Potential solutions (individuals) of optimization have been represented as a population of strings called chromosomes. In general, these solutions are encoded in the form of 0 and 1. Randomly, chromosomes have been generated as initial population and the process of evolution continued as generations. In each generation, population has been selected by employing selection methods and fitness value for every individual of the population is calculated. Depending upon these fitness values and replacement strategy, a set of chromosomes are selected as part of new population. And for some chromosomes crossover and/or mutation operations have been applied to generate new offspring to add to new population. This new population is evaluated during coming iterations. This will continue until maximum number of iterations or good predefined fitness value has been reached. The best chromosome at successful end of this process may be the optimal solution to the given problem [26, 27].

In this paper, GA has been adopted for selection of optimal features. The given each patients information has been encoded as a chromosome. Randomly, initial population has been generated for experimentally determined population size. The fitness of each chromosome has been calculated by using the following fitness function specified in Eq. (12.2),

$$\text{fit}(c) = \text{acc}(c) + (\text{af} * b1) + (\text{pf} * b2) \tag{12.2}$$

where c represents the chromosome for which fitness value has been calculated. Here, C4.5 decision tree algorithm has been applied as part of fitness calculation. "acc" represents the prediction accuracy of disease by considering the features presented in chromosome "c". This has been obtained using C4.5 algorithm. "af" and "pf" represent number of features absent (not considered) and present (considered) in the chromosome "c". And $b1$ and $b2$ are balancing factors which are experimentally calculated constants. These balance factors give weightage for prominent and not prominent features in given ratio, since each attribute may contribute in decision making. The next generation has been produced by employing roulette wheel selection procedure and by applying single point crossover. The steps in GA have been presented below.

Algorithm:

Step 1: Represented the problem variable domain as a chromosome of fixed size. And initial population of size N, maximum number of iterations max_iter, the balancing constants b1, and b2 have been defined.

Step 2: Randomly, N number of chromosomes has been selected as initial population.

Step 3: Repeat the following steps 4 and 5 for max_iter times

Step 4: Calculated the fitness of each individual by using Eq. (12.2)

Step 5:   Roulette wheel selection method has been applied and selected the new
          population as follows:

          (i)   Rank the chromosomes according to their fitness values
          (ii) Select two weak (chromosomes with least fitness value) chromosomes
          (iii) Perform single-point crossover on selected chromosomes and replace
                them with newly generated population
          (iv) Then apply bit string mutation on first newly generated chromosome

Step 6:   Selected the best chromosome with highest fitness value from the popula-
          tion. And the features present in this chromosome (patient's record) are the
          best features selected by this algorithm.

12.4   Artificial Neural Network-Based Classification

An artificial neural network (ANN) is a popular classification approach which is
a conceptual computational representation of human brain. As brain, an ANN is
network of interconnected artificial neurons which can be depicted as a graph of
neurons as vertices and interconnections as edges. According to topology, learning
methodology, and orientations of connections, there are different variants of ANNs.
Feedforward backpropagation neural network is one of most popular feedforward
ANNs [28, 29]. This algorithm has been employed because of its efficiency, ability
to find the weights in a reasonable amount of the time. This is a variant of gradient
search and uses least square optimality criterion. In this algorithm, calculation of
the gradient of error with reference to given inputs and their weights by propagating
the error backward through the network [30] play vital role. ANN can classify the
dataset quickly and is trained over given training data until a predefined threshold
has been reached. The backpropagation algorithm for training employed in this work
has been outlined as follows:

Step 1:   According to the number of selected features, the number of input nodes
          (n) has been defined and number of hidden nodes (m), one output node has
          been defined. To eliminate local minima, to break symmetry, and to avoid
          immediate saturation of the activation function small random initial weights
          have been selected.
Step 2:   For each patient record, repeat the steps from 3 to 5
Step 3:   Feedforward computation

          (i)   Hidden layer inputs are computed using Eq. (12.3) for each hidden
                node from $k = 1$ to m where w_{ij} indicates the weight assigned to the
                edge from kth input node to jth hidden node.

$$H_k = w_{1k} * i_1 + w_{2k} * i_2 + \cdots + w_{nk} * i_n + \theta_1 \qquad (12.3)$$

(ii) Then the output of each hidden layer is calculated by using the sigmoid function presented in (12.4) as activation function.

$$\text{Out}_{H_k} = \frac{1}{1 + e^{-H_k}} \tag{12.4}$$

(iii) The input to output node is calculated by using Eq. (12.5)

$$O_{\text{input}} = \sum_{k=1}^{m} w_k * \text{out}_{H_k} + \theta_2 \tag{12.5}$$

(iv) The output of the output node is calculated by using the sigmoid function presented in (12.6)

$$O_{\text{output}} = \frac{1}{1 + e^{-O_{\text{input}}}} \tag{12.6}$$

Step 4: Backpropagation

(i) Calculated the error difference between target output (T) and obtained output.

$$E = O_{\text{output}} * \left(1 - O_{\text{output}}\right) * \left(T - O_{\text{output}}\right) \tag{12.7}$$

(ii) For each hidden node k, computed error with respect to output layer. And for each input node k computed error with respect to hidden layer by using Eq. (12.8), where j represents nodes in next highest layer.

$$E_k = O_k * (1 - O_k) * \sum_{j} E_j * w_{kj} \tag{12.8}$$

Step 5: Weights updation

(i) Each weight in the network has been updated by using Eqs. (12.9) and (12.10) where w_{ij} indicated the edge weight from node in ith layer to node in jth layer.

$$\Delta w_{ij} = (l) * E_j * O_i \tag{12.9}$$

$$w_{ij} = w_{ij} + \Delta w_{ij} \tag{12.10}$$

(ii) Each bias θ_j (bias in jth layer) in the network has been updated by applying Eqs. (12.11) and (12.12) where l is the learning rate which has fixed experimentally.

$$\Delta\theta_j = (l) * E_j \tag{12.11}$$

$$\theta_j = \theta_j + \Delta\theta_j \tag{12.12}$$

Step 6: Repeated the steps from 2 to 5 until threshold value for the error met.

12.5 Experimental Results and Analysis

The proposed system has been experimented on two different medical datasets, namely diabetics and thyroid which are selected from UCI machine repository. The diabetic dataset contains 9 attributes (8 attributes + 1 class attribute), and the thyroid dataset contains 28 attributes (27 attributes + 1 class attribute). The results obtained on these medical datasets were analyzed with respect to not only prediction accuracy and also considered time taken to train the NN, the memory space taken by feature space. The optimal set of features obtained for diabetic dataset includes glucose, blood pressure, insulin, age, and pregnancies. And for thyroid dataset are age, thyroxine, query on thyroxine, antithyroid, sick, thyroid, T131, hypothyroid, hyperthyroid, psych, TSH, T3, and TT4. The results have proven that the proposed system producing good improvement over without FS. The results have analyzed in Table 12.1.

The experiments were carried out for different number of hidden nodes and various values of parameters in NN and FGANN. There is 62.5% size reduction, 12.8% learning time reduction for diabetic dataset and 66.7% size reduction and 33.98% learning time reduction for thyroid dataset. FS also has been carried out for different number of max_iter and obtained various optimal sets of features. The best optimal feature set has been selected and that has been given as input for NN classifier. The prediction accuracy for two datasets has been greatly improved after applying FS. This analysis has been presented various graphs with respect to memory space, learning time, prediction accuracy in Figs. 12.2, 12.3, and 12.4, respectively.

Table 12.1 Summarized result analysis

	Memory size of feature space (KB)		Learning time (s)		Prediction accuracy (%)		Number of features used in diagnosis	
	NN	FGANN	NN	FGANN	NN	FGANN	NN	FGANN
Diabetics data	24	15	114.255	14.644	68.7479	88.901	8	5
Thyroid data	57	38	43.09	14.644	73.452	98.01	27	13

Fig. 12.2 Graph depicting storage space required by datasets

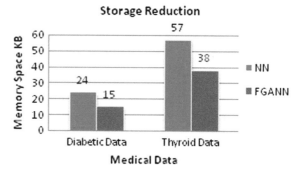

Fig. 12.3 Graph depicting learning time required by datasets

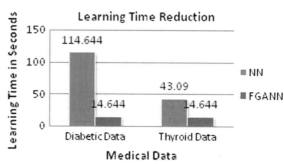

Fig. 12.4 Graph depicting learning time required by datasets

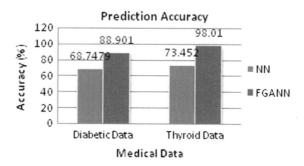

12.6 Conclusion

At a glance, the results seem to be very solid. Emerging information technologies, variations, and complexity involved in data, correlations of data driving toward more effective and efficient approaches in problem-solving and especially in healthcare analytics. These requirements demanding parallel processing mechanisms like human brain which has memory capability and efficient way of dealing with uncertainties. This work presented NN classifier for medical diagnosis by considering fuzzy GA to obtain prominent features. In the previous literature, the prediction accuracy was not greater than 85%, but classification accuracy in the proposed system is greater than 88% for diabetic and 98% for thyroid. In future, by increasing the number of hidden layers in NN the learning time and accuracy can be improved.

References

1. West D, Mangiameli P, Rampal R, West V (2005) Ensemble strategies for a medical diagnosis decision support system: a breast cancer diagnosis application. Eur J Oper Res 162:532–551
2. Gorunescu F, Belciug S (2016) Boosting back propagation algorithm by stimulus-sampling: application in computer-aided medical diagnosis. J Biomed Inform 63:74–81
3. Chandrashekar G, Sahin F (2014) A survey on feature selection methods. Comput Electr Eng 40:16–28
4. Rajeswari K, Vaithiyanathan V (2011) Fuzzy based modeling for diabetic decision support using artificial neural network. Int J Comput Sci Netw Secur 11(4):126–130
5. Schuerz M, Adlassnig K-P, Lagor C, Scheider B, Grabner G. Definition of fuzzy sets representing medical concepts and acquisition of fuzzy relationships between them by semi-automatic procedures
6. Kalpana M, Senthil Kumar AV (2011) Fuzzy expert system for diabetes using fuzzy verdict mechanism. Int J Adv Netw Appl 3(2):1128–1134
7. Hinton GE, Salakhutdinov RR (2006) Reducing the dimensionality of data with neural networks. Science 313:504–507
8. Guyon I, Elisseeff A (2003) An introduction to variable and feature selection. J Mach Learn Res 3:1157–1182
9. Ali Jan Ghasab M, Khamis S, Mohammad F, Jahani Fariman H (2015) Feature decision making ant colony optimization system for an automated recognition of plant species. Expert Syst Appl 42:2361–2370
10. Liu H, Yu L (2005) Toward integrating feature selection algorithms for classification and clustering. IEEE Trans Knowl Data Eng 17(4):491–502
11. Kohavi R, John GH (1997) Wrappers for feature subset selection. Artif Intell 97:273–324
12. Blum AL, Langley P (1997) Selection of relevant features and examples in machine learning. Artif Intell 97:245–270
13. Gletsos M, Mougiakakou SG, Matsopoulos GK, Nikita KS, Nikita AS, Kelekis D (2003) A computer-aided diagnostic system to characterize CT focal liver lesions: design and optimization of a neural network classifier. IEEE Trans Inf Technol Biomed 7(3):153–162
14. Sun Y-N, Horng M-H, Lin X-Z, Wang J-Y (1996) Ultrasound image analysis for liver diagnosis: a non invasive alternative to determine liver disease. IEEE Eng Med Biol Mag 93–101
15. Shilaskar S, Ghatol A (2013) Feature selection for medical diagnosis: evaluation for cardiovascular diseases. Expert Syst Appl 40:4146–4153
16. Goldberg D (1989) Genetic algorithms in search, optimization and machine learning. Addison-Wesley, Boston
17. Handels H, Rob Th, Kreusch J, Wolff HH, Pöppl SJ (1999) Feature selection for optimized skin tumour recognition using genetic algorithms. Artif Intell Med 16:283–297
18. Dhawan AP, Chitre Y, Kaiser-Bonasso C, Moskowitz M (1996) Analysis of mammographic microcalcifications using gray-level image structure features. IEEE Trans Med Imaging 15(3):246–259
19. Yamany SM, Khiani KJ, Farag AA (1997) Application of neural networks and genetic algorithms in the classification of endothelial cells. Pattern Recogn Lett 18:1205–1210
20. Yang J, Honavar V (1998) Feature subset selection using a genetic algorithm. IEEE Intell Syst Appl 13:44–49
21. Cybenko G (1989) Approximation by superpositions of a sigmoidal function. Math Control Signal 2(4):303–314

22. Paliwal M, Kumar UA (2009) Neural networks and statistical techniques: a review of applications. Expert Syst Appl 36(1):2–17
23. Piri S, Delen D, Liu T, Zolbanin HM (2017) A data analytics approach to building a clinical decision support system for diabetic retinopathy: developing and deploying a model ensemble. Decis Support Syst 101:12–27. https://doi.org/10.1016/j.dss.2017.05.012
24. Staub Selva et al (2015) Artificial neural network and agility. Procedia Soc Behav Sci 195:1477–1485
25. Jagadish HV, Gehrke J, Labrinidis A, Papakonstantinou Y, Patel JM, Ramakrishnan R, Shahabi C (2014) Big data and its technical challenges. Commun ACM 57:86–94
26. Ahmad F, Isa NAM, Hussain Z, Osman MK (2013) Intelligent medical disease diagnosis using improved hybrid genetic algorithm—multilayer perceptron network. J Med Syst 37:9934
27. Saxena A, Saad A (2007) Evolving an artificial neural network classifier for condition monitoring of rotating mechanical systems. Appl Soft Comput 7(1):441–454
28. Zorman M, Podgorelec V, Lenič M, Povalej P, Kokol P, Tapajner A (2003) Inteligentni sistemi in profesionalni vsakdan. CIMRŠ Univerze v Mariboru
29. Rajasekaran S, Vijayalakshmi Pai GA (2007) Neural networks, fuzzy logic, and genetic algorithms synthesis and applications. Prentice Hall of India, New Delhi
30. Amma NGB (2012) Cardiovascular disease prediction system using genetic algorithm and neural network. In: IEEE international conference on computing, communication and applications
31. Fasanghari M, Montazer GA (2010) Design and implementation of fuzzy expert system for Tehran Stock Exchange portfolio recommendation. Expert Syst Appl 37:6138–6147
32. Palfy M, Papez J (2007) Diagnosis of carpal tunnel syndrome from thermal images using artificial neural networks. In: Twentieth IEEE international symposium on computer-based medical systems (CBMS'07)

Printed in the United States
By Bookmasters